Professional Food Manager

Professional Food Manager

Editor

Oliva Cabral

scitus
academics

Professional Food Manager

Edited by **Oliva Cabral**

ISBN: 978-1-68117-553-9
Library of Congress Control Number: 2016942961

SCITUS Academics LLC,
www.scitusacademics.com
Box No. 4766, 616 Corporate Way,
Suite 2, Valley Cottage,
NY 10989

Printed in United States of America on Acid Free Paper

Preface

In order to manage your risk effectively, food service managers must learn the basic principles of food safety; use the information on the task every day. Food managers are responsible for the operation of a food service department, such as a grocery store, cafeteria or restaurant. Food managers are the frontrunners in the front and in restaurants of every kind in the kitchen. Responsible for the overall operation of the restaurant, food service managers hire stock, buy food and manage staff, and ensure that everyone is trained in proper food preparation, kitchen safety and health standards. Food managers make sure that the company is the right way and that the standards of the company are maintained. Food service managers are responsible for performing administrative, employee management and customer service duties. Administrative tasks often include inventory management, merchandising, cost control and product quality, signage, invoicing and stores inspection. Customer service duties may include promoting the company's image to employees and customers externally, resolving customer complaints to their satisfaction and responding to correspondence. It can also involve various contact responsibilities for customers, including helping customers with questions, providing samples and proposing products.

Professional Food Manager is specifically designed to students dealing with course in culinary and hospitality sector as well as for professionals in order to ensure the continued successful implementation of best practices in the field of food safety at work. The book highlights obstacles in the management of food safety and provides key insights into how to avoid them.

Table of Contents

Chapter 1

Global Hospitality Managers: Myth Or Reality?

ABSTRACT

Over the past forty years, the international management studies have expressed considerable interest in what has come to be known as 'best-practices' (Boxall and Purcell, 2000, 2003). The concept of best-practices suggests standardisation and homogenisation of the organisation's human resources through the employment of universally applicable managerial practices. This idea has been extremely appealing for MNCs during the 1980s and 1990s especially in industries like hospitality and tourism where pressures for standardisation due to the rapid global expansion were enormous. There are arguments however that it is difficult to imagine, that a single practice or set of practices would emerge as 'best' in any sense of the word, particularly in globalised organisations (Thomas, 2008). Moreover, a growing number of IHRM studies argue that 'a best practice is not best unless it incorporates contextual elements in its application' (Von Glinow et al., 2005, p.398). Thus, the dynamic and complex nature of the management function in global business today and the realisation that what works effectively in one country may not be as efficient in another, has led management scholars and practicing managers in continuous efforts to enhance their understanding of this context and its effects on international (hotel) managers. This is sought through the systematic study and exploration of management across cultures (cross cultural management), and international human resource management.

KEYWORDS

Hospitality Industry, Managerial Work, Cross-Cultural Management, ISHRM.

1. INTRODUCTION

"Until recently the dominance of American management theory led to the belief that 'one size fits all', that a good manager in the US will also be a good manager in other countries and that effective U.S. management practices will be effective anywhere. This view is now being supplanted with the knowledge that managerial attitudes, values, behaviours, and efficacy differ across national cultures. There is no one best way to manage a business. Differences in national cultures call for differences in management practices." (Newman and Nollen, 1996, p.753)

The rise of globalization has triggered a considerable interest in what has come to be known as 'best-practices' (Boxall and Purcell, 2000, 2003) from an international management studies perspective. The concept of best-practices suggests standardisation and homogenisation of the organisation's human resources through the employment of universally applicable managerial practices; this idea has been extremely appealing for hospitality multinational companies (MNCs) during the 1980s and 1990s. The high expatriate failure rates and the scarcity of managerial resources, alongside with the rise of awareness of the local context effects in managerial work, has resulted to a new direction in international managerial work research, largely affected by the 'global-local' question. This paper explores the wider implications of globalisation in hospitality managerial work and the level of convergence in managerial work practices that has been achieved so far.

1.1 The international hospitality manager

Hospitality as a modern phenomenon was shaped after World War Two and is closely linked with the development of mass tourism and the rapid growth of the airline industry. The hospitality and tourism industries together are the largest and fastest growing industry in the world (Clarke and Chen, 2007). The World Travel and Tourism Council (WTTC, 2009) estimated that hospitality and tourism as a global economy are directly and indirectly responsible for 11% of gross domestic product, 200 million jobs, 8% of total employment and 5.5 million new jobs per year until 2010.

The early adoption of internationalisation in the hotel industry came initially from U.S. hotel companies, who took the lead and moved across borders for supply and demand reasons; that was to satisfy the needs of American travellers as other trades internationalised (Nickson, 1998; Thompson et al., 1998). Since the early 1980s a growing number of these original American operators were acquired by U.K. based companies, and simultaneously other European and Asian companies began to compete on more broadly international scales (Thompson et al., 1998; Segal-Horn, 2000). The transformation of the sector in a truly globalised industry occurred after a prolonged period of mergers and acquisitions during the 1990s. This brought changes in the adopted growth strategies of many hotel companies who were now competing on a global basis (Price, 1993). While the hotel industry was traditionally dominated by individually and/or family owned properties it can be argued that it is heavily influenced by international/multinational hotel companies defined by Peng and Litteljohn (1997) as:

"multi-unit service organizations in which units operate under a system of decision-making permitting coherent policies and a common strategy through one or more decision-making centres, and where hotel units and corporate functions are linked to add value to each other by ownership or contractual relationships." (cited in Litteljohn, 2003, p.15)

In this globalised environment, the development of international hospitality managers is seen as being of critical importance for hospitality MNCs. Thus, since the early 1990s, the rapid growth of international hotel chains and its effects on managerial work, have drawn the attention of researchers (i.e. Gilatis and Guerrier, 1994; Nebel et al., 1995; Gilbert and Guerrier, 1997; Ladkin and Juwaheer, 2000). The personality characteristics required of the international hotel managers include people and interpersonal skills, adaptability, flexibility and tolerance, cultural sensitivity and intercultural competence followed by emotional maturity, industry experience, and self-confidence (Gilatis and Guerrier, 1994; Shay and Tracey, 1997; Feng and Pearson, 1999; Kriegl, 2000). International etiquette, demonstrating an understanding of international business matters, the ability to work with limited resources and effectively manage stress were judged to be relatively important, while functional and technical skills were rated as the lowest priority for managers. Research also indicates that in an international hospitality organisation building managers' cross cultural skills may be far harder but more important than developing their functional and technical skills (Gilatis and Guerrier, 1994; Shay and Tracey, 1997; Kriegl, 2000). This is because of the high level of interpersonal and relational skills required where the host country culture and the needs of a diverse customer base must be understood and catered for. The knowledge and competences of GMs are wide-ranging and include not only the enabling capabilities (Leonard-Barton, 1995) essential for survival within the international hotel industry but also the supplemental and core capabilities specific to companies' market positions and strategies, and competitive advantage. International hospitality managers are seen as change agents who help corporations to cope with the fast changing environment. In addition, international hotel chains such as Starwood, Hilton, Intercontinental and Accor invest in the development of a cohort of international managers by using staff from both the host and parent countries they operate (Nickson, 1998; Jones et al., 1998). The recruitment and selection of this management 'cadre' is conducted in assessment centres and requires a variety of aspects such as a mix of competencies, technical skills, strong personality, cultural sensitivity and adaptability (D'Annunzio-Green, 1997). Teare (1995) provides a wide – ranging set of issues that arise from the internationalisation of the industry; the education training and development of managers, is one of the main six issues reviewed. In addition, Kriegl (2000, p.64) suggests that international hospitality operations' success 'depends largely on the availability of qualified managers who are able to export, translate and maintain their companies' operational standards and service consistency overseas'.

Despite the economic significance and global spread of the international hospitality industry, the majority of hospitality management literature reflects what has happened in the US and the UK since the early 1980s. Only recently have studies focused on what is happening in the rest of the Europe or the world. This was made possible through the contribution of a steadily growing number of overseas students in the U.S. and U.K. business schools who deliver hospitality programs. The most popular forms of research used to study the hospitality industry outside the Anglo-American context, is the use of country case studies (i.e. Kim, 1994; Christou, 1999; Agut et al, 2003) and studies within the context of the international hospitality business (D'Annunzio-Green, 1997). Despite the relatively slow progress, hospitality research persistently reflects the Anglo-American universalist approach to management. Thus, it can be argued that the changes currently taking place in international hospitality management can be better understood under a cross-cultural management perspective.

2. MANAGING IN DIFFERENT CULTURAL CONTEXTS

The extremely high failure of U.S. expatriate managers in non-western countries assignment has driven the creation of research focused on national cultures and its effects in organisational context (i.e. Hofstede, 1980, 1991; Tayeb, 1988, 1994; Trompenaars, 1993). Several attempts have been made to conceptualise and measure differences in cultures among nations, and to relate cultural differences to differences in management practices. This is sought through the systematic study and exploration of management across cultures (cross cultural management); although cross-cultural management is often regarded as a discipline of international management, is not a clearly demarcated discipline of management (Soderberg and Holden, 2002). Yet for many management scholars the term is meaningful because it implies (a) procedures and policies relating to the management of workforces with different cultural backgrounds, and (b) moderating the impact of cultural differences in the execution of management tasks (ibid, p.103). Adler (1991) argues that cross cultural management studies the behaviour of people and organisations in different countries and cultures around the world; he also suggests that "cross cultural management expands the scope of domestic management and encompass the international and multicultural spheres" (ibid., p.11). The most well known examples include the international survey results reported in Hofstede (1980, 1991), Laurent (1983) and Trompenaars (1993). This body of research clearly indicates that the national culture interpretation and adaptation are a prerequisite to the comparative understanding of national management practice. Triandis (1982) observed that specific management actions could be facilitated or inhibited by culturally determined orientations. He made a distinction between Dionysian cultures, where subordinates are motivated through close interpersonal affiliation and Apollonian cultures, where there the relationship between manager and subordinate is characterised by tasks and formality. For Newman and Nollen (1996) national culture is seen as a central organising principle of employees' understanding of work, their approach to it, and the way in which they expect to be treated.

National culture implies that one way of acting or one set of outcomes is preferable to another. Thus, when management practices are inconsistent with these deeply held values, employees are likely to feel dissatisfied, distracted uncomfortable and uncommitted. As a result, they may be less able or willing to perform well. House et al. (2002), in their GLOBE study, separated aspects of culture into its 'etic' and 'emic' qualities. The former are common for all cultures and the latter are culture-specific. The authors believe that these qualities can be used to explain similarities and differences in organisational practices and leadership behaviours. In particular, culture has been shown to shape the individual's perceptions and behaviours towards job design, supervision and rewards (Aycan et al., 2000; Smith et al., 2001). Furthermore, national cultures differ in the degree to which they enable managers to adopt non-homegrown practices (Tayeb, 1995), although recent research suggests that some contingency-type theories may be applicable (Ralston et al., 1999). Empirical evidence indicates that a variety of management practices differ by national culture, including decision making (Schneider and De Meyer, 1991), strategy (Ayoun and Moreo, 2008), leadership style (Dorfman and Howell, 1988; Puffer, 1993), and human resources management (Luthans et al., 1993).

The main obstacle in the effort to understand these management differences across different countries and cultural settings is that researchers and practitioners must often rely on theory originally developed in the United States. This has resulted to the fact that management theory development has been synonymous largely with what happens in this country and its MNCs operating around the globe are responsible for the dissemination of this US-centric management knowledge. Arguably the most vocal proponent of this position, Guest has suggested that HRM is 'American, optimistic, apparently humanistic and also superficially simple' (Guest, 1990, p.379), and has argued that HRM can be seen as a contemporary manifestation of the American dream owing to its emphases on the potential for human growth, the desire to improve opportunities for people at work, and the role of strong leadership reinforced by strong organisational culture (Guest

1990). In that respect, many management scholars argue that international and cross-cultural management studies originate in the individualistic achievement-oriented management culture of the United States (i.e. Doktor et al, 1991; Jackson, 2002; Tayeb, 2005; Thomas, 2008). This body of US-centric literature also suggests (Ruigrok and van Tulder, 1995; Marschan, 1996; Mayrhofer and Brewster, 1996; Ralston et al., 1997; Ferner and Quintanilla, 1998) that where a nation has a strong and distinctive culture this carries over into the nation's organisations, the most cited examples being Japanese, German and United States' organisations. Although MNCs operating independent of national borders continue to have their assets, sales, work-force, ownership and control highly concentrated in the country where their corporate headquarters are located (Ferner and Quintanilla, 1998). Thus, universality in international (hospitality) management theory, research and practice is accepted without question (Kiessling and Harvey, 2005). It can be argued however that attempts to establish a common corporate culture in an international or multi-national firm can be undermined by the strength of national cultures, and research points to the fact that the cultures of individual countries are both more stable and more powerful than those of individual organisations (Newman and Nollen, 1996; Hirst and Thompson, 1996; Pauly and Reich, 1997). Furthermore, the concept that human beings are a resource to further the executive ends of an organisation is a concept that is contrary to the values of many non-Western cultures (Thomas, 2008). Perhaps in its most instrumental conceptualisation, this may also be contrary to the values of many 'Western' cultures. Little thought is given to the implications of the underlying concepts in people management theory, nor to its manifestations in the policies and practices that multinational corporations employ across different countries (Ferner and Quintanilla, 1998). Boyacigiller and Alder (1991) argue that this bias in theory development is the result not of an inherent belief in the superiority of U.S. management but of parochialism – a lack of awareness of alternative contexts, models, research and values. As a result culture is often ignored in management research and universality is assumed (Thomas, 2008). Recently, a number of scholars have challenged the assumption that management theories formulated in one country are applicable universally, and have demonstrated in their research that management theories applicable in one country cannot be generalised directly to other countries (i.e. House et al., 2004; Javidan et al., 2005). Yet, progress in developing new context relevant theory in international management has been slow (Peterson, 2004; Tsui, 2004).

2.1 ISHRM and the global-local question in managerial work

The identification of the complexities associated with managing people from different cultural contexts when a company pursues business across national borders, has been a major aspect of IHRM research (Cullen, 1999; Evans et al, 2002; Schneider and Barsoux, 2003; Dowling and Welch, 2004). IHRM is an area of research and practice that is embedded in international management, which is in turn embedded in the broad field of international business (De Cieri et al., 2007). Peltonen (2006, p.523) defines IHRM as 'a branch of management studies that investigates the design and effects of organizational human resource practices in cross-cultural contexts', and Welch (1994, p.162) advocates that 'IHRM involves moving people around the globe'. The core difference between HRM and IHRM is premised in the fact that HRM is relevant within a single country, while IHRM seeks to explore added complexity due to a diversity of national contexts of operation and to the inclusion of different national categories of workers (Dowling, 1999; Evans et al., 2002; Schuler et al., 2002).

Increasing attention has focused on the strategic nature of IHRM and the implications of strategy for organisational performance (De Cieri and Dowling, 2006). Schuler et al, (1993, p.422) define strategic IHRM as 'human resource management issues, functions, and policies and practices that result from the strategic activities of multinational enterprises and that impact the international concerns and goals of those enterprises'. The present level in the evolutionary process of human resource management, which complicates the human resource functions and systems that much more, is strategic global human resource management (SGHRM). Viewing the human resource functions in an SGHRM context exacerbates the international human resource issues due to the co-ordination efforts required to implement the strategy of the corporation (Kiessling and Harvey, 2005). Thus, from a SGHRM perspective, in today's global business environment, global organisations must utilise all possible sources of competitive advantage, and human assets are one of these sources (Barney, 1991; Schuler and Rogovsky, 1998). Many of the debates on the strategic nature of international and global HRM have covered similar territory to those in domestic SHRM that is whether organisations should primarily build upon their internal assets or upon assessments of the external environment to develop competitive advantage (Bartlett and Ghoshal, 1989; Prahalad and Hamel, 1990). MNCs typically face the conundrum of 'the dual imperatives of local responsiveness and global integration' (Taylor et al., 1996, p.962). While the former strategy allows firms to customise goods and services to better suit local laws, customs and consumer tastes the latter achieves cost reductions through economies of scale and scope (Hill, 2006, pp.395-400). Such policies are tempered by the characteristics of the industrial sector within which the MNC operates and/or the type of product or service that it produces. Similarly, the implementation of new technologies and production practices within international subsidiaries often requires the export of new management and work practices from the parent firm to the host country subsidiary. Managing people consistently within and between subsidiaries across the world suggests that companies will evaluate whether to adopt local practices and simultaneously adopt global practices (Brewster, 2001, 2006; Edwards and Kuruvilla, 2005).

The above management dilemmas and challenges have not been new; since the early 1980s a critical question has been, whether differences among management practices and national cultures matter to managerial performance. The 'global-local' question originates from the wider cultural 'convergence-divergence' debate, which is premised on the assumption that in given enough time, cultures will converge to the point that no difference in values, attitudes, beliefs and behaviour exist (Kerr et al, 1960; Perlmutter, 1969; Eisenhardt, 1973; Dunphy, 1987; Inglehart and Beker, 2000). As a result, societies and organisations will become similar due to the universal application of management practices (Dowling et al, 1999; Schneider and Barsoux, 1997, 2003; Brewster, 2006). On the other hand, divergent theorists argue that in today's complex globalised business environment, international companies must manage, on a day-to-day basis, the national institutional and cultural differences presented in various localities, and this highlights another form of balance for enterprises where they have to achieve a judicious and effective equilibrium between some local HRM practices and elements of consistency across their subsidiaries (Brewster, 2006). While the adoption of global, 'culture-free' (convergence) or local 'culture-specific' (divergence) managerial practices may represent polar extremes, a third approach appears as a unifying model. In this respect, there are writers who argue that corporate culture can influence, but not eliminate, national culture values (Ricks et al., 1990; Adler, 1997), or who argue for 'crossvergence', which is defined as a combination of organisational culture and national culture values (Ralston et al., 1993). Basically, crossvergence theorists argue that as the global economy grows, countries will influence one another economically as well as culturally; in that respect crossvergence means that different management approaches are expected to converge in the middle (Vance and Paik, 2006). Regardless of whether management is converging, diverging or crossverging, culture can be seen as having an important influence on performance (Fisher and Hartel, 2003).

2.2 One problem, three different approaches

The solutions to the culture problem identified in international management literature are respectively to reflect, eliminate or adapt. In that respect, top management is a key factor in determining the overall international strategy and its implementation approach (Hax, 1989; Roth, 1995). Their belief in the generalisability or specificity of policies and practices is a driving force of MNC's choices (Heenan and Perlmutter, 1979; Hedlund, 1986). There are three distinctive practices within SIHRM in relation to this decision: the 'best-practice' approach; the 'best-fit' approach; and the 'resource-based view' (RBV).

The 'best-practice' argues for a universal paradigm, which promotes the idea of convergence across countries, and sees the main aim of SIHRM to improve organisational performance (Delaney and Huselid, 1996). Taylor et al. (1996) have labelled this as 'exportive' SIHRM orientation where the parent firms HRM system is being transferred to its different affiliates. Its arguments are based primarily on surveys of 'leading edge' companies. It appears as the dominant managerial practice within the United States of America but is also widely used in the UK, Ireland, Australia, New Zealand and Northern Europe (Brewster, 2001, 2006). The existing definitions extend from arguments that best practice approaches should ensure that minimum legislative requirements are met (Price, 1994) to those that proffer utopian views of people management approaches achieving increased levels of productivity, performance and profits (Pfeffer, 1998; Huselid, 1995). Boxall and Purcell (2003) argue that there is little overall agreement amongst the proponents of a single best practice vision; instead a range of practices are noted by researchers including participation, empowerment, job rotation, training and skill development, and teams (Osterman, 1994; Pfeffer, 1994). Furthermore, when trying to establish models of best practices it is increasingly difficult to find agreement amongst authors on which practices to deploy (Purcell, 1999; Marchington and Grugulis, 2000). The variety of models has often been classified together as 'bundles' of HR practices known as 'High Performance Work Practices' (HPWPs), which are particularly popular in the US literature (Harris et al., 2003). They are based upon the idea that to outperform their competitors firms must follow a high commitment management (HCM) 'recipe' for people management. The various 'best-practice' models have faced forceful critique by many authors on the grounds of the empirical research approach and techniques applied, the absence of consistency on what constitutes 'best-practice(s)' and the limits of diffusion of these supposed panaceas (Marchington and Grugulis, 2000; Purcell, 2001). Additional criticism is based on the proposition that there are widespread benefits from the adoption best practice; in reality not everyone gains from this universal approach (Marchington and Grugulis, 2000; Purcell, 2001). The claims made in the name of best practice tend to belie in the U.S. (Boxall and Purcell, 2000), when they take place at the same time as growing income inequalities and further efforts to weaken labour markets through the use of outsourcing, temps and sub-contracting. Where some of the working population no doubt benefit from these best practices, the rest (a majority) are exposed to some of the harsher realities of modern day capitalism (Marchington and Grugulis, 2000). Boxall and Purcell (2000) argue that there is overwhelming evidence against a set of universal HR practices, based on the existence of contextual factors such as national and international variations in culture, market regulation and traditions of management. Thus, it can be argued that the ignorance of the managerial work context poses as one of the main weakness in the adoption of best practices in different cultural contexts.

On the other hand, Morden (1995) argues that international managers and multinational companies may need to take a 'best-fit' or 'contingency' approach to any issues related with the management function; he also suggests (ibid.) that the style of management must be appropriate to the prevailing local contingencies. Taylor et al. (1996) have labelled this as an adaptive orientation, in which each subsidiary or affiliate firm develops its own HRM system, reflecting the local environment. In short, this approach suggests that firms should be able to take advantage of cultural diversity to help them differentiate their products and services and satisfy customers and workforces, whilst at the same time minimising the effects of cultural diversity where standardisation is prioritised. In terms of the development of key human resources this balance is seen in the use of career structures for talented managers regardless of nationality and the use of host and third country managers in senior positions (Adler and Ghadar, 1990; Paauwe and Dewe, 1995; Scullion and Paauwe, 2004). It can be argued that the notion of 'fit' is fundamental to all contingent SHRM models, as Beer et al. (1984 p.13 in Bratton and Gold, 1999, p.48) suggest "there must be a 'fit' between competitive strategy and internal HRM strategy and a fit among the elements of the HRM strategy". Typically these are identified as 'external fit' between the organisation's competitive strategy or stage of development, and its HRM approaches, and 'internal fit' between the individual HRM practices and policies adopted (Schuler and Jackson, 1987; Baird and Meshoulam, 1988; Sanz-Valle et al., 1999). The idea that HR practices should be driven by the organisation's competitive strategy has raised major criticisms (Boxall and Purcell, 2000). Firstly, such a model fails to recognise the need to align employee interests with the organisation or comply with prevailing social norms and legal requirements in the course of this process (i.e. Boxall, 1996, Lees, 1997; Schein, 1997). Despite the employers' definite superior bargaining power, the managers' role in firms (among others) is to meet the baseline needs of employees whose skills are crucial for the organisation's survival (Coff, 1997; Boxall, 1998). This is especially so in highly competitive labour markets such as in hotels and retail, that are based on the economies of replication or 'known organisational routines' (Purcell, 1995). Secondly there is an absence of an effective SHRM fit theory due to the tendency of authors to base their SHRM arguments on implicit assumptions. Wright and Sherman (1999) identify these implicit assumptions as the belief that a particular business strategy requires a particular set of responses from employees and that a specific set of HRM practices produces a specific set of responses from employees. Thirdly, there is the problem of an inconsistent use of 'fit' models (Wright and Sherman, 1999; Boxall and Purcell, 2000). Fourthly, Wright and Sherman (1999) refer to problems associated with inconsistent constructs of HRM that is HRM practices, HR skills and HR behaviours, when describing types of 'fit'. Finally it is apparent that there are methodological and empirical issues associated with investigating 'fit'. Research by Miller (1992) for example, suggests that competitive strategy is often multi-dimensional and subject to important variations across industries. It is suggested that these 'research driven' problems need to be addressed to aid theoretical development of SHRM (Wright and Sherman, 1999). In summary, best-fit or contingency models argue that HR strategy becomes more effective, when it is designed to fit certain critical contingencies in the organisation's specific context. While many criticisms have been levelled at the 'best-fit' SHRM approach it is not completely impotent (Boxall and Purcell, 2000). Instead it is more useful to suggest that the 'best-fit' approach require further embellishments, including the wider consideration of the internal factors and mechanisms, which influence strategy and HRM practices over time.

Finally, the 'resource-based view' (RBV) or 'integrative' SIHRM orientation, attempts to take the 'the best' HRM approaches and use them throughout the organisation in the creation of a worldwide system (Taylor et al., 1996). This suggests that certain groups of human resources (i.e. managers) achieve a privileged status within organisations due to their capacity to transfer tacit knowledge to new markets and provide sustainable competitive advantage (Prahalad and Hamel, 1990; Scullion and Starkey, 2000; Jacob, 2003). Thus, organisations must develop a cadre of managers who have a global mindset as a way of thinking within the global marketplace (Kedia and Mukherji, 1999; Paul, 2000; Begley and Boyd, 2003). These managers must develop a pluralistic management perspective that encourages and maintains multiple perspectives in order to solve complex global problems (Aguirre, 1997; Reynolds, 1997; Harvey et al., 1999). Torrington et al. (2005) argue that this model sits very well with the increasing attention being given to the notion of 'human capital' (i.e. Boxall, 1996; Edvinsson and Malone, 1997) where it is the collective nature and quality of the people in the organisation, which provide the potential for future competitive advantage. The majority of RBV models suggest that a firm's resources should be 'valuable, rare, inimitable and non-substitutable' (VRIN) to achieve a competitive leading edge (Barney, 1991; Chadwick and Cappelli, 1999; Boxall and Purcell, 2003;

Morris et al., 2006). In the first instance value is where a resource allows a firm to improve its market position relative to its rivals so human resources can be valuable due to the range of competencies and skills individuals possess (Wright et al., 1994). The capacity of a resource to be rare is associated with the limited availability of a resource relative to demand (Barney, 1991; Hoopes et al., 2003). Consequently the scarcity of people with the requisite skills and behavioural capacities in the labour market, despite wider levels of unemployment, suggest that human resources potentially fulfill the criteria of being rare resources (Marchington and Wilkinson, 2002). The third principle of inimitability is particularly important to the RBV and emerges where resources are difficult or costly to reproduce so competitors will not be able to mimic the company's advantage (Barney, 1991; Boxall and Purcell, 2003; Hoopes, et al., 2003). The potential barriers to imitation are conceived via the combination of 'social complexity, causal ambiguity and unique historical circumstances' (Boxall and Purcell, 2000, p.194). Social complexity may exist where human resources are involved in interdependent relationships, inside and outside the organisation (Boxall and Purcell, 2003). Causal ambiguity is argued to occur because it is often impossible to establish how successful a firm is and what has led to its success (Barney, 1991). The final inimitability barrier – 'unique historical circumstances' – is concerned with the difficulties of learning for new entrants and rivals in the market (Prahalad and Hamel, 1990; Wright et al., 1994; Boxall and Purcell, 2003). The fourth characteristic of non-substitutability identifies a resource cannot be obsolete or traded for another factor (Barney, 1991; Wright et al., 1994; Boxall and Purcell, 2000; Marchington and Wilkinson, 2002). Human resources satisfy this condition, as they cannot become outmoded, and are transferable across a range of markets, products and technologies. A fifth characteristic has also been identified for resources capable of achieving sustained competitive advantage (Grant, 1991; Kamoche, 1996) through the use of the economic term 'rent' (an above average rate of return in competitive markets), where resources and capabilities should also retain the value they have fashioned for their own on-going advantage. The capacity of human resources to do this is easily seen in the importance many organisations attach to retaining particular groups of staff and the opportunities to create additional benefits for human resources where a firm has outperformed its rivals.

Despite the fact that the RBV is much less developed in the literature on SHRM compared with the debate between 'best-practice' and 'best-fit' (Torrington et al., 2005), its attractiveness to the SHRM literature has been widely discussed (Barney et al., 2001; Marchington and Wilkinson, 2002; Morris et al., 2006). For a theory which emphasises the value of people to the strategic success of the organisation it seems unlikely to claim many opponents from the SHRM discipline. However, there are arguments regarding the limitations of RBV, focused on this approach's overemphasis on organisational differentiation and competitive advantage (Oliver, 1997; Kamoche, 2001; Marchington and Wilkinson, 2002; Boxall and Purcell, 2003). Furthermore, Oliver (1997) argues that while forces for differentiation exist, the forces for similarity should not be discounted. Such forces include external coercion from legislation, normative traditions in reaching customers and dealing with suppliers and routes to achieving efficiencies. Probably most prominent are the arguments of the institutional perspective, which identify examples of isomorphism as part of the forces of similarity (Boxall and Purcell, 2003; Bjorkman, 2006; Rosenzweig, 2006). Isomorphism is a process, which constrains one actor (company) in a constituency to resemble the other actors when all are facing the same environmental circumstances (DiMaggio and Powell, 1983; Paauwe and Boselie, 2003). Isomorphism concepts suggest common factors for institutional assimilation should be recognised where the RBV and 'best fit' SHRM approaches (through the strategic models) promote companies' unlimited opportunities to distinguish themselves competitively. According to Warner and Joynt (2002), convergence of management structures and practices in organisations across cultures is likely to be driven by factors such as:

- The fast diffusion of management practices, for example, through international management education and activities by MNCs (Parker, 1998) which could be described with normative and mimetic isomorphism (Di Maggio and Powell, 1983).

- The 'uninversal language' of technology which determines organisational structures and work processes (Hickson and Pugh, 1995).

3. DISCUSSION

From the above discussion on the 'global-local' question and the different SIHRM orientations ('best-fit', 'best-practice', RBV), it can be argued that the existing models do not fully answer some of the criticisms that have been levelled against the fields of international (hospitality) management and SIHRM (Brewster et al., 2005, De Cieri et al., 2007). It can be argued that hospitality organisations operating in local, national or international level are too complex to be able to establish exactly that one aspect of resource management, however, valuable, rare, inimitable and non-substitutable it might be, provides the definitive route to better financial performance. It is more valuable, from researchers' and practitioners' perspective, to understand how, why and under which conditions organisations develop what they perceive to be 'strategic managerial resources'.

It is also important to acknowledge that management is socially constructed and cannot be separated from context (Dierdorff et al., 2009). Of the various criticisms levelled at studies of managerial work (Mintzberg, 1973; Willmott, 1984; Martinko and Gardner, 1985; Hales and Nightingale, 1986; Stewart, 1989; Hales, 1999), one of the most insistent has been the reluctance to situate either models of managerial work or empirical evidence on managerial activity within a broader contextual (i.e. institutional and/or cultural) framework (Hofstede, 1980; Boyacigiller and Adler, 1991; Doktor et al., 1991; Hales and Tamangani, 1996; Lewis, 2000; Hofstede, 2001). Armstrong's (1986, p.19) observation that 'the orthodox management literature is neutered by its own studied ignorance of the context and purpose of management activity', is also echoed by other writers (i.e. Marples, 1967; Burgoyne and Hodgson, 1984; Hales, 1986; Willmott, 1987;

Stewart, 1989; Tsoukas, 1994). Consequently, in much of the SIHRM literature, the political, social, economic, cultural and institutional contexts are treated as simple contingency factors - this despite the fact that it is known that at firm level globalisation is a country specific phenomenon (Ferner and Quintanilla, 1998). Thus, SIHRM appears to be reproducing the flawed universalist assumptions of the broader fields of international management and international business (Westwood, 2006). The narrow focus of studies in this field is best expressed through the persistence of researchers to study expatriates; this fact leads to the question of 'who or what has remained ignored?' (De Cieri et al., 2007). To this extend, there are still more questions to be answered such as 'are expatriates performing better than local managers within the same sector/organisation?' or 'how similar or different managerial work is when comparing expatriates with local managers?'

Moreover, SIHRM models tend to confuse cross-national with cross-cultural differences, which risks confusing what will remain stable and is hard for firms to influence with what terms might be able to standardise with judicious management (Brewster et al., 2005). The issue of cultural relativity has tempted researchers and practitioners to focus on the 'hard' or 'core' HR functional processes (Easterby-Smith et al., 1995). The field tends to ignore the subtle ways in which cultural/national differences influence the experienced reality of HRM (Earley and Singh, 2000). In addition, the inadequate conceptualisation of national effects, results in culture being used as an unsatisfactory 'catch-all' for national differences (Edwards and Kuruvilla, 2005). The wider convergence-divergence debate also tends to assume that the HRM system as a whole has to converge or remain divergent, rather than considering whether some parts of the overall HR system might be converging, in some regions or geographies, while other parts might be diverging (Brewster et al., 2005). Moreover, even within a single HR function there might be convergence at one level but divergence at another. An HR function operates at multiple levels, including philosophy, policy, programme, practice and process (Schuler, 1992). However, by contrast, an over-emphasis on comparative factors risks freezing the discourse in terms of national differences. Any analysis of IHRM must consider the range of distinctive national and local solutions to HRM issues with which the firm has to deal, the strategic pressures that make these national models more receptive to change and development, and the firm-level processes through which such change and development in actual HRM practice is delivered (Sparrow and Hiltrop, 1997).

Finally, Kiessling and Harvey (2005) argue that research into IHRM is increasingly difficult to pursue, and complicated, as culture and its alignment with strategy may be very difficult to research through strictly quantitative methods. Quantitative methods sometimes do not work well in the study of international management due to the complexity and unstructuredness of the problems, with multiple important interactive relationships that cannot be studied in a quick or easy fashion (Wright, 1996). Thus a growing number of researchers (i.e. Boyacigiller and Adler, 1991; Daniels, 1991; Brewer, 1992; Parkhe, 1993; Kiessling and Harvey, 2005) argue that a combination of both quantitative and qualitative research approaches is needed in order to advance the field, and direct such study towards reconstructing the 'how' of IHRM (Pieper, 1990). Moreover, the growing influence of the interpretivist's approach to management studies have led a growing number of scholars to challenge the assumption that management theories formulated in one country are applicable universally, and have demonstrated in their research that management theories applicable in one country cannot be generalised directly to other countries (i.e. House et al., 2004; Javidan et al., 2005). Based on this argument, researchers indicate that managers working in a global context might experience disconnection between international management theory prescriptions and the imperatives of the local context unless international management theory is grounded in the realities of the local context (i.e. Horwitz et al., 2002; Ngambi, 2004). Qualitative methodologies provide researchers with a valuable tool in this regard because they assume a value-laden approach in terms of data that are collected and interpreted within the context in which it is generated (Kriek et al., 2009). Dennis and Garfield (2003, p.297) assert 'quantitative research is theory in search of data while qualitative research is data in search of theory'. Qualitative field studies provide rich data and a strong base on which to build theory; this view is supported by a number of authors (i.e. Eisenhardt, 1989; Strauss and Corbin, 1990; Mendenhall et al., 1993; Miles and Huberman, 1994; Cassel, et al., 2006) who have written extensively on the use of qualitative research methods in building theory. Others assert qualitative research precedes theoretical and empirical research in driving a field to maturity (Weick, 1989; Van de Ven, 1989). The use of qualitative methods, such as case studies (Eisenhardt, 1989), has long been noted to be especially important in the progression of theoretical knowledge in organisational behaviour studies (Gummesson, 2000; Denzin and Lincoln, 2003; Yin, 2003). Werner (2002) found that case studies and other qualitative methodologies have still limited application in international (hospitality) management studies. It is not clear though, whether this is because of the inherent difficulties of qualitative research (Wright et al., 1988), because of the dominant philosophical position of the quantitative method superiority, or because of some other reason (Werner, 2002).

4. CONCLUSION

In summary, the above discussion has indicated a clear need for a change in the direction of international hospitality research on managerial work and the HRM interventions used for its managerial resources. Although the question 'is it possible to create a truly global hospitality management cadre?' seems to remain unanswered, there are signs both for convergence and divergence in international hospitality management practices. What needs to be done, is to see more national and cross national 'U.S. and U.K. independent' research that will help us develop a deeper understanding of the interaction between management and the local context. Moreover, integration between cross-cultural management, SIHRM and other international management sub-fields is also needed in order to create a common research language in international management studies.

REFERENCES

1. Adler, N. (1997) International dimensions of organizational behavior, 3rd Edition, Cincinnati, OH: South-Western College Publishing.

2. Adler, N. J. & Ghadar, F. (1990) Strategic Human Resource Management: A Global Perspective. In R. Pieper (Ed.) Human Resource Management in International Comparison, Berlin/New York: De Gruyter, pp. 235-260.

3. Adler, N.J. (1991) International Dimensions of Organizational Behavior, Belmont, CA: Wadsworth Publishing.

4. Aguirre, M. (1997) Multiculturalism in a Labor Market with Integrated Economies, Management Decision, 35(7), 489–96.

5. Agut, S., Grau, R., & Peiro, J.M. (2003) Competency needs among managers from Spanish hotels and restaurants and their training demands, Hospitality Management, 22, 281-295.

6. Armstrong, P. (1986) Management control strategies and inter-professional competition: the cases of accountancy and personnel management. In D. Knights & H. Willmott (Eds) Managing the labour Process, Aldershot: Gower, pp.19-43.

7. Aycan, Z., Kanungo, R.N., Mendonca, M., Yu, K., Deller, J., Stahl, G. & Khursid, A. (2000) Impact of culture on human resource management practices: A ten country comparison, Applied Psychology: An International Review, 49(1), 192-220.

8. Ayoun, B. & Moreo, P. (2008) Does national culture affect hotel managers' approach to business strategy? International Journal of Contemporary Hospitality Management, 20(1), 7-18.

9. Baird, L. & Meshoulam, I. (1988) Managing two fits of strategic human resource management, Academy of Management Review, 13(1), 116-128.

10. Barney, J. (1991) Firm Resources and Sustained Competitive Advantage, Journal of Management, 17, 99–120.

11. Barney, J.B. Wright, M. & Ketchen, D.J. Jr (2001) The resource-based view of the firm: Ten years after 1991, Journal of Management, 27, 625-641.

12. Bartlett, C.A. & Ghoshal, S. (1989) Managing across boarders, London: Hutchinson.

13. Begley, T. & Boyd, D. (2003) The Need for a Corporate Global Mind-Set, Sloan Management Review, 44(2), 78–86.

14. Bjorkman, I. (2006) International human resource management research and institutional theory. In G.K., Stahl & I. Bjorkman (Eds.) Handbook of Research in International Human Resource Management Cheltenham: Edward Elgar, 36-48.

15. Boxall, P. (1996) The strategic HRM debate and the resource-based view of the firm, Human Resource Management Journal, 6, (3), 59-75.

16. Boxall, P. (1998) Achieving competitive advantage through human resource strategy: towards a theory of industry dynamics, Human Resource Management Review, 8(3), 265-288.

17. Boxall, P. & Purcell, J. (2000) Strategic human resource management: where have we come from and where should we be going? International Journal of Management Reviews, 2(2), 183-203.

18. Boxall, P. & Purcell, J. (2003) Strategy and Human Resource Management, Basingstoke: Palgrave Macmillan.

19. Boyacigiller, N.A. & Alder, N.J. (1991) The parochial dinosaur: Organisational science in a global context, Academy of Management Review, 16(2), 262-290.

20. Bratton, J. & Gold, J. (1999) Human Resource Management: Theory and Practice, London: Macmillan.

21. Brewer, T.L. (1992) An Issue-area Approach to the Analysis of MNE-Government Relations, Journal of International Business Studies, 23(2), 295–309.

22. Brewster, C. (2001) Human Resource Practices in Multinational Companies. In M.J. Gannon, & K. Newman, (Eds.) Handbook of Cross-Cultural Management, New York: Blackwell, 126-141.

23. Brewster, C. (2006) Comparing HRM policies and practices across geographical borders. In G.K. Stahl & I. Bjorkman (Eds.) Handbook of Research in International Human Resource Management, Cheltenham: Edward Elgar, 68-90.

24. Brewster, C., Sparrow, P. and Harris, H. (2005) Towards a new model of globalizing HRM, Int. Journal of Human Resource Management, Vol.16, No.6, pp. 949-970.

25. Burgoyne, J.G. & Hodgson, V. (1984) An experiential approach to understanding managerial action. In J.G., Hunt, D., Hosking, C.A., Schriesheim, & R. Stewart, (Eds) Leaders and Managers: International Perspectives on Managerial Behavior and Leadership, New York: Pergamon.

26. Cassell, C., Symon, G., Beuhring, A. & Johnson, P. (2006) The role and status of qualitative methods in management research: An empirical account, Management Decision, 44(2), 290-303.

27. Chadwick, C and Cappelli, P. (1999) Alternatives to Generic Strategy Typologies in Strategic Human Resource Management. In P.M., Wright, L.D., Dyer, Boudreau and G.T., Milkovich (Eds) Research in Personnel and Human Resource Management: Strategic Human Resources Management in the Twenty-First Century - Supplement 4, London: JAI Press.

28. Christou, E. (1999) Hospitality management education in Greece: An exploratory study, Tourism Management, 20, 683-691.

29. Clarke, A. & Chen, W. (2007) International Hospitality Management Concepts and Cases, Oxford, Butterworth-Heinemann.

30. Coff, R. (1997) Human assets and management dilemmas: coping with hazards on the road to resource based theory, Academy of Management Review, 22(2), 374-402.

31. Cullen, J.B. (1999) Multinational Management: A strategic approach, Cincinnati: South-Western College Publishing.

32. D' Annunzio-Green, N. (1997) Developing international managers in the hospitality industry, International Journal of Contemporary Hospitality Management, 9(5/6), 199-208.

33. Daniels, J.D. (1991) Relevance in International Business Research: A Need for More Linkages, Journal of International Business Studies, 22(2), 177–86.

34. De Cieri, H. & Dowling, P.J. (2006) Strategic human resource management in multinational enterprises: developments and directions. In G.K. Stahl, & I. Björkman, (Eds) Handbook of Research in International Human Resource Management, Cheltenham: Edward Elgar, 15–35.

35. De Cieri, H., Cox, J.W. & Fenwick, M. (2007) A review of international human resource management: Integration, interrogation, imitation, International Journal of Management Reviews, 9(4), 281–302.

36. Delaney, J. & Huselid, M. (1996) The impact of human resource management practices on perceptions of organisational performance, Academy of Management Journal, 39(4), 949-969.

37. Dennis, A.R. & Garfield, M. (2003) The adoption and use of GSS in project teams: Towards more participative processes and outcomes, MIS Quarterly, 27(2), 289-317.

38. Denzin, N.K. & Lincoln, Y.S. (2003) The Landscape of Qualitative Research, Thousand Oaks: Sage.

39. Di Maggio, P. & Powell, W. (1983) The iron cage revisited: Institutional isomorphism and collective rationality in organizational fields, American Sociological Review, 48, 147-160.

40. Dierdorff, E., Rubin, R. & Morgeson, F. (2009) The Milieu of Managerial Work: An Integrative Framework Linking Work Context to Role Requirements, Journal of Applied Psychology, 94(4), 972–988.

41. Doktor, R.H., Tung, R. & Von Glinow, M. (1991) Incorporating international dimensions in management theory building, Academy of Management Review, 16(2), 259-261.

42. Dorfman, P. W. & J. P. Howell (1988) Dimensions of National Culture and Effective Leadership Patterns: Hofstede revisited, Advances in International Comparative Management, 3, 127-150.

43. Dowling, P.J. (1999) Completing the puzzle: issues in the development of the field of international human resource management, Management International Review, 39(3), 27-43.

44. Dowling, P.J. & Welch, D.E. (2004) International Human Resource Management: managing people in a multinational context, 4th Edition, London: Thomson Learning.

45. Dowling, P.J., Welch, D.E. & Schuler, R.S. (1999) International Human Resource Management: Managing People in a Multinational Context, 3rdedition, Cincinnati: South-Western College Publishing.

46. Dunphy, D. (1987) Convergence/Divergence: a temporal review of the Japanese enterprise and its management, Academy of Management Review, 12, 445-459.

47. Earley, P.C. & Singh, H. (eds) (2000) Innovations in International and Cross-Cultural Management, Thousand Oaks, CA, Sage.

48. Easterby-Smith, M., Malines, D. & Yuan, L. (1995) How Culture-Sensitive is HRM? A Comparative Analysis of Practice in Chinese and UK Companies, International Journal of HRM, 6(1), 31-59.

49. Edvinsson, L. & Malone, M.S. (1997) Intellectual Capital: Realizing your Company's True Value by Finding Its Hidden Brainpower, New York: Harper Business.

50. Edwards, T. & Kuruvilla, S. (2005) International HRM: national business systems, organizational politics and the international division of labour in MNCs, International Journal of Human Resource Management, 16(1), 1–21.

51. Eisenhardt, K.M. (1989) Building theories from case study research, Academy of Management Review, 14(4), 532-550.

52. Eisenhardt, S.N. (1973) Tradition, change and modernity, New York: John Wiley & Sons.

53. Evans, P., Pucik, V. & Barsoux, J.L. (2002) The Global Challenge: Frameworks for International Human Resource Management, New York: McGraw-Hill/Irwin.

54. Feng, F. & Pearson, T.E. (1999) Hotel Expatriates in China: selection criteria, important skills and knowledge, repatriation concerns, and causes of failure, International Journal of Hospitality Management, 18(2), 309-321.

55. Ferner, A. & Quintanilla J. (1998) Multinationals, national business systems and HRM: the enduring influence of national identity or a process of 'Anglo-Saxonization', The International Journal of Human Resources Management, 9(4), 710-731.

56. Fisher, G. & Hartel, C. (2003) Cross-cultural effectiveness of Western expatriate-Thai client interactions: lessons learned for IHRM research and theory, Cross Cultural Management, Vol.16, No.1, pp.1-21.

57. Gilatis, N. & Guerrier, Y. (1994) Managing international career moves in international hotel companies. In C.P., Cooper, & A. Lockwood, (Eds.) Progress in Tourism, Recreation and Hospitality Management, 5, 229-241.

58. Gilbert D. & Guerrier Y. (1997) **UK hospitality managers past and present**, Service Industries Journal, 17, 115-132.

59. Grant, R. (1991) A resource-based theory of competitive advantage: implications for strategy formulation, California Management Review, Spring, 114-135.

60. Guest, D. (1990) Human resource management and the American Dream, Journal of Management Studies, 27(4), 377-97.

61. Gummesson, E. (2000) Qualitative Methods in Management Research, London: Sage.

62. Hales, C. (1999) Why do managers do what they do? Reconciling evidence and theory in accounts of managerial work, British Journal of Management, 10, 335-350.

63. Hales, C. and Nightingale, M. (1986) What are the unit managers supposed to do? A contingent methodology for investigating managerial role requirements, International Journal of Hospitality Management, 5(1), 3-11.

64. Hales, C. & Tamangani, Z. (1996) An investigation of the relationship between organisational structure, managerial role expectations and managers' work activities, Journal of Management Studies, 33, 731-756.

65. Harris, H., Brewster, C. & Sparrow, P. (2003) International Human Resource Management, London: CIPD.

66. Harvey, M., Novicevic, M. and Speier, C. (1999) The Impact of Emerging Markets on Staffing the Global Organizations, Journal of International Management, 5(2), 34–46.

67. Hax, A.C. (1989) Building the Firm of the Future, Sloan Management Review, Spring, 75–82.

68. Hedlund, G. (1986) The hypermodern MNC – A heterarchy? Human Resource Management, 25(1), 9–35.

69. Heenan, D.A. & Perlmutter, H.V. (1979) Multinational organization development, Reading, MA: Addison–Wesley.

70. Hickson, D. & Pugh, D. (1995) Management Worldwide, London: Penguin Books.

71. Hill, C. (2005) Global Business Today, 4th Edition, Boston: McGraw-Hill Irvin.

72. Hirst, P. & Thompson, G. (1996) Globalization in Question, Cambridge: Polity Press.

73. Hofstede, G. (1980) Cultures' consequences: international differences in work-related values, Beverly Hills: Sage.

74. Hofstede, G. (1991) Cultures and organizations: software of the mind, London: McGraw Hill.

75. Hofstede, G. (2001) Culture's Consequences: Comparing Values, Behaviors, Institutions and Organizations across Nations, 2nd Edition, Thousand Oaks, CA: Sage.

76. Hoopes, D.G. Madsen, T.L. & Walker, G. (2003) Guest editors' introduction to the special issue: Why is there a resource-based view? Towards a theory of competitive heterogeneity, Strategic Management Journal, 24, 889-902.

77. Horowitz, F.M., Kamoche, K. & Chew I.K.H. (2002) Looking east: Diffusing high performance work practices in the Southern Afro-Asian context, International Journal of Human Resource Management, 13(7), 1019-1041.

78. House, R., Javidan, M., Hanges, P. & Dorfman, P. (2002) Understanding Cultures and implicit leadership theories across the globe: an introduction to project GLOBE, Journal of World Business, 37, 3-10.

79. House, R.J., Hanges, P.J., Javidan, M., Dorfman, P. & Gupta, V. (Eds.) (2004) Leadership, Culture and Organisations: The Globe Study of 62 Societies, Thousand Oaks, CA: Sage.

80. Huselid, M. (1995) The impact of human resource management practices on turnover, productivity and corporate financial performance, Academy of Management Journal, 38(3), 635-672.

81. Inglehart R. & Beker, W.E. (2000) Modernization, cultural change, and the persistence of traditional values, American Sociological Review, 65, 19-51.

82. Jackson, T. (2002) International HRM: a cross-cultural approach, London: Sage.

83. Jacob, N. (2003) Intercultural management, London: Kogan Page.

84. Javidan, M., Stahl, G.S., Brodbeck, J. & Wilderom, C.P.M. (2005) Cross-border transfer of knowledge: Cultural lessons from Project Globe, Academy of Management Executive, 19(2), 59-76.

85. Kamoche, K. (1996) Strategic human resource management within a resource-capability view of the firm, Journal of Management Studies, 33(2), 213-233.

86. Kamoche, K.N. (2001) Understanding Human Resource Management, Buckingham: Open University Press.

87. Kedia, B. & Mukherji, A. (1999) Global Managers: Developing a Mindset for Global Competitiveness, Journal of World Business, Vol.34, No.3, pp.230–251.

88. Kerr, C., Dunlop, J.T., Harbison, F.H. & Myers, C.A. (1960) Industrialism and industrial man: the problems of labour and management in economic growth, London: Heinemann.

89. Kiessling, T. & Harvey, M. (2005) Strategic Global Human Resource Management Research in the 21st Century: The Role of the Mixed-Method Research Methodology, International Journal of Human Resource Management, 16(1), 22-45.

90. Kim, S.M. (1994) Tourist hotel general managers in Korea: a profile, International Journal of Hospitality Management, 13(1), 7-17.

91. Kriegl, U. (2000) International hospitality management, Cornell Hotel and Restaurant Administration Quarterly, 41(2), 64-71.

92. Kriek, H.S., Beaty, D. & Nkomo, S. (2009) Theory building trends in international management research: An archival review of preferred methods, South African Journal of Economic and Management Sciences, 12(1), 126-135.

93. Ladkin, A. & Juwaheer, D.T. (2000) The career paths of hotel general managers in Mauritius, International Journal of Contemporary Hospitality Management, 12(2), 119-125.

94. Laurent, A. (1983) The cultural diversity of Western Corporations of Management, International Studies of Management and Organisations, 13(1), 75-96.

95. Lees, S. (1997) HRM and the legitimancy market, International Journal of Human Resource Management, 8(3), 226-243.

96. Leonard-Barton, D. (1995) Wellsprings of Knowledge: Building and Sustaining the Sources of Innovation, Boston, Masachusetts: Harvard Business School Press.

97. Lewis, R.D. (2000) When Cultures Collide, London: Nicholas Brealey.

98. Litteljohn, D. (2003) Hotels. In B. Brotherton, (Ed.) The International Hospitality Industry: Structure, Characteristics and Issues, Oxford: Butterworth-Heinemann, 5-29.

99. Luthans, F., Welsh, D.H.B, & Rosenkrantz, S.A. (1993) What do Russian managers really do? An observational study with comparisons to US managers, Journal of International Business Studies, 24(4), 741-761.

100. Marchington, M. & Grugulis, I. (2000) 'Best practice' human resource management: perfect opportunity or dangerous illusion, International Journal of Human Resource Management, 11(6), 1104-1124.

101. Marchington, M. & Wilkinson, A. (2002) People Management and Development at Work: Human Resource Management at Work, 2nd Edition, London: CIPD.

102. Marples, D.L. (1967) Studies of managers - a fresh start? Journal of Management Studies, 4(2), 282-299.

103. Marschan, R. (1996) New structural forms and inter unit communication in multinationals, Helsinki: Helsinki School of Economics.

104. Martinko, M. & Gardner, W. (1985) Beyond Structured Observation: Methodological Issues and New Directions, Academy of Management Review, 10(4), 676-695.

105. Mayrhofer, W. & Brewster, C. (1996) In praise of ethnocentricity: expatriate policies in European multinationals, International Executive, 38(6), 749-778.

106. Mendenhall, M., Beaty, D.T. & Oddu, G. (1993) Where have all the theorists gone? An archival review of the international management literature, International Journal of Management, 10(2), 146-153.

107. Miles, M. & Huberman, M. (1994) Qualitative Data Analysis, Thousand Oaks, CA: Sage.

108. Miller, D. (1992) The generic strategy trap, Journal of Business Strategy, 13(1), 37-42.

109. Mintzberg, H. (1973) The nature of Managerial Work, New York: Harper & Row.

110. Morden, A.R. (1995) National Culture and the Culture of the Organisation, Cross Cultural Management, 2(2), 3-12.

111. Morris, S.S., Snell, S.A. & Wright, P.M. (2006) A resource-based view of international human resources: towards a framework of integrative and creative capabilities. In Stahl G.K. and I., Bjorkman (Eds.) Handbook of Research in International Human Resource Management Cheltenham: Edward Elgar, 433-448.

112. Nebel, E.C., Lee, J. & Vidakovic, B. (1995) Hotel general manager career paths in the United States, International Journal of Hospitality Management, 14(3/4), 245-260.

113. Nickson, D. (1998) A Review of Hotel Internationalisation with a Particular Focus on the Key Role Played by American Organisations, Progress in Tourism and Hospitality Research, 4, 53-66.

114. Oliver, C. (1997) Sustainable Competitive Advantage: Combining Institutional and Resource-based views, Strategic Management Journal, 18(9), 697-17.

115. Osterman, P. (1994) How common is workplace transformation and who adopts it? Industrial and Labor Relations Review, 26(1), 46-67.

116. Paauwe, J. & Boselie, P. (2003) Challenging 'strategic HRM' and the relevance of the institutional setting, Human Resource Management Journal, 13(3), 56-70.

117. Paauwe, J. & Dewe, P. (1995) Human resource management in multinational corporations: theories and models. In Harzing, A. & Van Ruysseveldt, J. (eds.) International Human Resource Management, London: Sage, 75-98.

118. Parker (1998) Globalization and business practice: Managing across borders, London: Sage.

119. Parkhe, A. (1993)'Messy' Research, Methodological Predisposition, and Theory Development in International Joint Ventures, Academy of Management Review, 18(2), 227–268.

120. Paul, H. (2000) Creating a Global Mindset, Thunderbird International Business Review, 42(2), 187–200.

121. Pauly, L. & Reich, S. (1997) National Structures and Multinational Corporate Behavior: Enduring Differences in the Age of Globalization, International Organization, 51(1), 1–30.

122. Peltonen, T. (2006) Critical theoretical perspectives on international human resource management. In G.K., Stahl & I. Björkman (Eds) Handbook of Research in International Human Resource Management, Cheltenham: Edward Elgar, 523–535.

123. Perlmutter, H.V. (1969) The Tortuous Evolution of the Multinational Company, Columbia Journal of World Business, Jan/Feb, 9-18.

124. Peterson, R.B. (2004) Empirical research in international management: A critique and future agenda. In R. Marschan-Piekkari & C. Welch (Eds.) Handbook of Qualitative Research Methods for International Business, Cheltenham: Edward Elgar, 25-55.

125. Pfeffer, J. (1994) Competitive Advantage through People, Boston, MA: Harvard Business School Press.

126. Pfeffer, J. (1998) Six Dangerous Myths About Pay, Harvard Business Review, May-June, 109-119.

127. Pieper, R. (ed.) (1990) Human Resource Management: An International Comparison, Berlin: De Gruyter.

128. Prahalad, C. & Hamel, G. (1990) The Core Competence of the Corporation, Harvard Business Review, 68(3), 79-91.

129. *Price, L. (1994) Poor personnel practice in the hotel and catering* industry: does it matter? Human Resource Management Journal, 4(4), 44-62.

130. Price, S. (1993) Strategic groups and foreign investment in the International Hotel Industry, MBA dissertation, University of Bradford, UK.

131. Puffer, S.M. (1993) A riddle wrapped in an enigma: Demystifying Russian managerial motivation, European Management Journal, 11(4), 473-480.

132. Purcell, J. (1995) Corporate strategy and its link with human resource management strategy. In J. Storey (Ed.) Human Resource Management: a critical text, London: Routledge, 63-86.

133. Purcell, J. (1999) 'Best Practice' and 'Best Fit': chimera or cul-de-sac? Human Resource Management Journal, 9(3), 26-41.

134. Purcell, J. (2001) The Meaning of Strategy in HRM. In J. Storey (Ed.) Human Resource Management: a critical text, 2nd Edition London: Int. Thomson Business Press, 59-77.

135. Ralston, D., Gustafson, D., Cheung, F. & Terpstsra, R. (1993) Differences in managerial values: A study of US, Hong Kong and PRC managers, Journal of International Business Studies, 24, 249 – 275.

136. Ralston, D.A., Egri, C.P., Stewart, S., Terpstra, R.H. & Kaicheng, Y. (1999) Doing Business in the 21st Century with the New Generation

of Chinese Managers: A Study of Generational Shifts in Work Values in China, Journal of International Business Studies, 30(2), 415–428.

137. Ralston, D.A., Holt, D.H., Terpstra, R.H. & Yu, K. (1997) The impact of national culture and economic ideology on managerial work values: A study of the United States, Russia, Japan, and China, Journal of International Business Studies, 28(1), 177-207.

138. Reynolds, C. (1997) Strategic Employment of Third Country Nationals: Keys to Sustaining the Transformation of HR Functions, Human Resource Planning, Vol.20, No.1, pp.33–40

139. Ricks, D., Toyne, B, & Martinez, Z. (1990) Recent developments in international management research, Journal of Management, 16, 219 – 253.

140. Rosenzweig, P.M. (2006) The dual logics behind international human resource management: pressures for global integration and local responsiveness. In G.K., Stahl & I. Bjorkman (Eds.) Handbook of Research in International Human Resource Management, Cheltenham: Edward Elgar, 36-48.

141. Roth, K. (1995) Managing international interdependence: CEO characteristics in a resource-based framework, Academy of Management Journal, 38, 200–231.

142. Ruigrok, W. & van Tulder, R. (1995) The Logic of International Restructuring, London: Routledge.

143. Sanz-Valle, R., Sabater-Sanchez, R. & Aragon-Sanchez, A. (1999) Human Resource management and business strategy links: an empirical study, International Journal of Human Resource Management, 10(4), 655-671.

144. Schein, E. (1997) Increasing organisational effectiveness through better human resource planning and development, Sloan Management Review, 19(1), 1-20.

145. Schneider, S.C. & Barsoux, J. (Eds.) (2003) Managing Across Cultures, 2nd Edition, London: FT-Prentice Hall

146. Schneider, S.C. & Barsoux, J. (Eds) (1997) Managing Across Cultures, London: Prentice Hall.

147. Schneider, S.C. & De Meyer, A. (1991) Interpreting and responding to strategic issues: The impact of national culture, Strategic Management Journal, 12, 307-320.

148. Schuler, R. & Jackson, S. (1987) Linking competitive strategy and human resource management practices, Academy of Management Executive, 3(1), 207-219.

149. Schuler, R.S. (1992) Linking the People with the Strategic Needs of the Business, Organizational Dynamics, Summer, 18-32.

150. Schuler, R.S. & Rogovsky, N. (1998) Understanding Compensation Practice Variations Across Firms: The Impact of National Culture, Journal of International Business Studies, 29(1), 159–177.

151. Schuler, R.S., Budhwar, P.S. & Florkowski, G.W. (2002) International human resource management: review and critique, International Journal of Management Reviews, 4(1), 41–70.

152. Schuler, R.S., Dowling, J. & De Cieri, H. (1993) An integrative framework of strategic international human resource management, Journal of Management, 19(2), 419-459.

153. Scullion, H. & Paauwe, J. (2004) International Human Resource Management: Recent Developments in Theory and Empirical Research, In A., Harzing & J., Van Ruysseveldt (Eds.) International Human Resource Management, 2nd Edition, London: Sage, 65-88.

154. Scullion, H. & Starkey, K. (2000) In search of the changing role of the corporate human resource function in the international firm, International Journal of Human Resource Management, 11(6), 1061-1081.

155. Segal-Horn, S. (1994) Are Service Industries going Global? In C. Armistead (Ed.) The Future of Services Management, The Cranfield Management Research Series, London: Kogan Page, 41-63.

156. Shay, J. and Tracey, J.B. (1997) Expatriate Managers: Reasons for failure and implications for training, Cornell Hotel and Restaurant Administration Quarterly, February, Vol. 38, No.1, pp.30-35.

157. Smith, P.B., Fisher, R. & Sale, N. (2001) Cross-cultural industrialisation and organizational psychology. In C.L. Cooper & I.T. Robertson (Eds) International Review of Industrial and Organizational Psychology, 16, 147–194.

158. Søderberg, A.M. & Holden N. (2002) Rethinking Cross Cultural Management in a Globalizing Business World, International Journal of Cross Cultural Management, 2(1), 103-121.

159. Sparrow. P.R. & Hiltrop, J.M. (1997) Redefining the Field of European Human Resource Management: A Battle between National Mindsets and Forces of Business Transition, Human Resource Management, 36(2), 1-19.

160. Stewart, R. (1989) Studies of managerial jobs and behaviour: the ways forward, Journal of Management Studies, 26(1), 1-10.

161. Strauss, A.L. & Corbin, J. (1990) Basics of Qualitative Research: Grounded Theory Procedures and Techniques, Newbury Park, CA: Sage.

162. Tayeb, M. (1994) Organisations and National Culture: Methodology considered, Organisational Studies, 15(3), 429-446.

163. Tayeb, M. (1988) Organizations & national culture: A comparative analysis, London: Sage.

164. Tayeb, M. (1995) Organizations and National Culture: Methodology Considered, Organizational Studies, 15(6), 429–446.

165. Tayeb, M. (2005) International Human Resource Management: a multinational company perspective, Oxford: Oxford University Press.

166. Taylor, S., Beechler, S. & Napier, N. (1996) Toward an integrative model of strategic international human resource management, Academy of Management Review, 21(4), 959–985.

167. Teare, R. (1995) The International Hospitality Business: a thematic perspective, International Journal of Contemporary Hospitality Management, 7(7), 55-73.

168. Thomas, D.C. (2008) Cross-Cultural Management: Essential Concepts, 2nd Edition, London: Sage.

169. Thompson, P., Nickson, D., Wallace, T. & Jones, C. (1998) Internationalisation and Integration: A Comparison of Manufacturing and Service Firms, Competition and Change, 3, 387-415.

170. Torrington, D., Hall, L. & Taylor. S. (2005) Human Resource Management, 6th Edition, London: FT-Prentice Hall.

171. Triandis, HC. (1982) Culture in organisational theories, International Studies of Management and Organisation, 12(4), 139–169.

172. Trompenaars, F. (1993) Riding the waves of culture, London: Nicholas Brealy.

173. Tsoukas, H. (1994) What is management? An outline of a metatheory, British Journal of Management, 5(4), 289-301.

174. Tsui, A.S. (2004) Contributing to global management knowledge: A case for high quality indigenous research, Asia Pacific Journal Of Management, 21(4), 491-513.

175. Van de Ven, A.H. (1989) Nothing is quite so practical as a good theory, Academy of Management Review, 14(4), 486-9.

176. Vance, C. & Paik, Y. (2006) Managing a global workforce: challenges and opportunities in international human resource management, New York: M.E. Sharpe Inc.

177. Warner, M. & Joynt, P. (2002) Managing across cultures: issues and perspectives, 2nd Edition, London: Thomson.

178. Weick, K.E. (1989) Theory construction as disciplined imagination, Academy of Academy of Management Review, 14(4), 516-531.

179. Welch, D. (1994) Determinants of International Human Resource Management Approaches and Activities: A suggested framework, Journal of Management Studies Vol. 31, No. 2, pp.139-164.

180. Werner, S. (2002) Recent Developments in International Management Research: A Review of 20 Top Management Journals, Journal of Management, 28(3), pp. 277-305.

181. Westwood, R. (2006) International business and management studies as an orientalist discourse: a postcolonial critique, Critical Perspectives on International Business, 2, 91–113.

182. Willmott, H. (1984) Images and ideals of managerial work: a critical examination of conceptual and empirical accounts, Journal of Management Studies, 21(3), 349-368.

183. Willmott, H. (1987) Studying managerial work: a critique and a proposal, Journal of Management Studies, 24(3), 249-270.

184. Wright, L.L. (1996) Qualitative International Management Research. In B.J., Punnett & O. Shenkar (Eds) Handbook for International Management Research, Cambridge, MA: Blackwell, 63–81.

185. Wright, L.L., Lane, H.W., & Beamish, P.W. (1988) International management research: Lessons from the field, International Studies of Management, International Studies of Management and Organization, 18(3), 55–71.

186. Wright, P.M. & Sherman, W.S. (1999) Failing to find fit in Strategic Human Resource Management: Theoretical and Empirical Problems. In P.M., Wright, L.D., Dyer, Boudreau & Milkovich, G.T. (Eds) Research in Personnel and Human Resource Management: Strategic Human Resources Management in the Twenty-First Century - Supplement 4, London: JAI Press, 53-74.

187. Wright, P.M. McMahan, G.C. & McWilliams, A. (1994) Human Resources and Sustained Competitive Advantage: A Resource-based perspective, International Journal of Human Resource Management, 5(2), 301-326.

188. World Travel & Tourism Council (2009) Progress and Priorities 2008-2009, annual report available from www.wttc.org, accessed: 11/10/2009

189. Yin, R.K. (2003) Case Study Research - Design and Methods, 3rd Edition, London: Sage.

Chapter 2

The Contribution Of The Entrepreneur-Manager Of Small Tourism Enterprises To The Success Of Internet Marketing Activities – The Greek Case

ABSTRACT

It is widely accepted that small enterprises constitute a catalyst for creating jobs and generally growing the economy. One industry that is characterized by the proliferation of small enterprises is the tourism industry. A critical factor for this industry reaching its full potential is, however, access to markets. The Internet is seen by many, as having the potential to facilitate small tourism businesses to access their markets, understand them better, expand their market, reach and serve their customers more effectively, regardless of their location.

The focus of this study is to identify those entrepreneur-manager factors that are present in a small tourism enterprise which influence the success with which the Internet can be used to market the business.

These factors are related to the entrepreneur-manager characteristics and not to the technology factors. The identification of those factors and the estimation of their impact on Internet marketing activities will allow small tourism businesses to access markets more readily.

The methodology of this study was based on a quantitative research that was conducted via the Internet. The measuring instrument was a questionnaire that was linked to a 7-point Likert scale. The sample was selected from a database of Greek tourism businesses.

The results of this study indicate that the level of involvement of entrepreneur-managers, their ability to select the suitable internet agency, their ability to brief satisfactory the latter one, their writing skills for communicating effectively their services, their understanding of the web 2.0 technologies as well as their marketing orientation are important to the successful use of the Internet for the marketing of small tourism businesses in Greece.

KEYWORDS

entrepreneur-manager, small tourism enterprises, internet, marketing, Greece

1. INTRODUCTION

The importance of the SME for the Greek Economy is quite substantial. Voulgaris et al. (2004) write that SMEs create 50,000 new jobs in Greece annually, and represent 60% of the employment in the manufacturing and service sector. Also Zimmerer et al. (2002) point to the high labour absorption capacity of small businesses suggesting that small businesses generate wealth at a faster rate than larger firms would. Also they point out; the growth in numbers of people employed by small businesses is greater than the growth in the contribution of these enterprises to GDP.

An important factor that distinguishes large businesses and the small ones is the substantial impact of the entrepreneur-manager to the day-to-day activity of the latter ones and to their decision making (Gilmore et al., 2001). According to Marcati et al. (2008) "entrepreneurs' innovativeness and personality play a key role in the adoption of innovations in small- and medium-sized Enterprises (SMEs)". Since the Greek tourism sector is particularly characterized by the proliferation of SMEs the study of the contribution of the entrepreneur-manager to the success of internet marketing activities is particularly relevant.

1.1 The Sme Tourist Businesses

The tourism and travel business is consisted of five sectors, which are the accommodation sector, the transport sector, the travel organizers sector, the attractions sectors and the destination organization sector (Middleton, 1993). More specifically each of the aforementioned sectors is analyzed as in Figure 1. The tourism and travel sector is mainly considered as a service industry. As Cooper et al. (1998) have stated "An understanding of the complexity of the service product concept is an essential prerequisite for successful marketing". According to Lovelock et al. (2007) services have distinctive marketing challenges in relation to those of goods.

More specifically:

- Most services cannot be inventoried
- The intangible elements of the service product usually dominate value creation
- Services are usually difficult to visualize and understand
- Customers may be involved in co-production

- People may be part of the service experience
- Operational inputs and outputs tend to vary more widely
- The time factor often assumes great importance
- Distribution may take place through nonphysical channels

Particularly for the tourist services, Middleton (ibid) has identified some additional distinctions, namely:

- The seasonality of the tourist services
- The interdependence of the tourist services e.g. staying in a hotel, visiting museums, using transportations.
- The high fixed cost of the tourist services production.

The above elements influence the tourism business performance and viability and "explain why much of the travel and tourism is considered to be a high risk business" (Middleton, ibid, p. 32). In this high risk business, entrepreneurs are asked to react accordingly to the rapidly changing market trends, making difficult, if necessary, adjustments in order to become more competitive. To this direction, the finding of inexpensive production factors and the use of them, in such a way as to render the maximum result, is a very crucial task (Karagiannis, 2004).

The tourist sector is quite important for the Greek economy since in 2004 the tourist income represented the 6.2% of the Greek GDP in relation to 3.5% that was in 1995, according to a study of Sampaniotis & Vorlow (2006).

For the purpose of the present research there will be a focus in the accommodation sector and particularly in the hotels sub-sector, since the latter one is quite important taking into consideration that there are about 10.000 hotels in Greece, according to the Hellenic Chamber of Hotels.

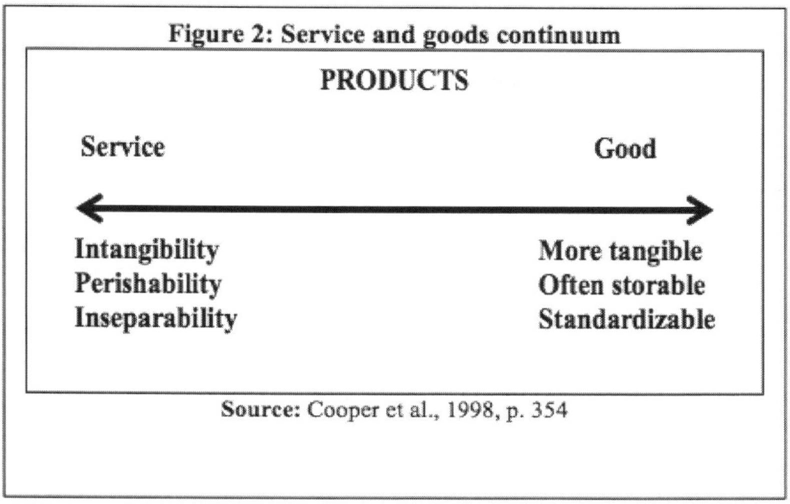

Figure 1. Service and goods continuum

1.2 Internet Marketing Activities

The Internet has been widely adopted globally. According to the Vorlow's study (2008: 9), the Internet penetration in 2004 was 83% in Holland, 79% in Sweden, 71% in Germany, and 67% in UK. The 57% of internet users conducts searches via search engines. Moreover 2 out of 5 internet users in UK, Finland and Luxemburg make use of online tourist services like reservations, payments, information.

Also the Internet has been proliferated considerably the last ten years and a lot of tools are at the disposal of the entrepreneurs-managers. Anderson (2009) cites that "There are many different ways to use the Internet as a marketing tool, including blogs, social and professional media, and search engine optimization (SEO). More specifically:

- Blogs: It is a kind of website maintained by an individual in order to place regular entries of commentary, or posting videos, photos. Entries are displayed in reverse chronological orders. The creator of the blog has special privileges, for example he can delete comments or even ban the participation of specific individuals if they do not confront to the blog's rules. There are literally millions of blogs written by all kinds of different people. A typical blog combines text, images and links to other blogs. Web pages and other media related to its topic. For an agency, a blog represents a very interesting way to have an ongoing dialogue with both customers and prospects. A distinctive feature of a blog is the ability for the reader to leave comments in an interactive format. This is an important part of many blogs because it allows conversation between the bloggers and their readers. Providing focused information on a particular topic along with the ability to create a conversation with those who read the post is what has made blogs popular.

- Social media and professional media: that includes Facebook, LinkedIn, Twitter, YouTube. More specifically:
  - Facebook: It is the most popular social networking media with over of 300 million users who are networked with their friends and share comments, videos, photos, groups etc. From a marketing perspective there are two aspects of Facebook that a company should pay attention to. First is the social networking aspect. There are an increasing number of companies that have created a Facebook fan page. There are a few companies that are making their Facebook fan pages an integral part of their Web sites and online presence. The second aspect of Facebook marketing is Facebook advertising, one of the most targeted forms of advertising currently available on the Internet. This provides a new way for companies to communicate and connect to new audiences. Because of the information that is provided in a Facebook user's profile, a company can target its ads to a very specific type of person. For example, you can target an ad to display to only those people who have a particular city listed in their profile.
  - LinkedIn: It is the Facebook for professionals. At the last count it had more than 42 million skilled professional users from all over

the world, comprising 150 different industries. When someone joins the site, he creates a personal profile that sums up their professional achievements. These profiles enhance his visibility to prospective clients, current and former colleagues and potential business partners. Members can add connections by inviting business contacts to join LinkedIn, and to connect to other members as a trusted contact.

- Twitter: Twitter is a "micro-blog." In 140 characters or less you can describe what you are doing, reading, thinking, or whatever else you think others will find of value. For most people Twitter has moved beyond describing what you are having for lunch. And it doesn't really matter whatever you care or not. But it might matter to certain clients. Twitter can represent another communication channel for customers who are inclined to be on the leading edge of technology and communication.

- YouTube: YouTube is a video sharing site where users can upload view and share video clips. Google acquired the site in November 2006. Before the launch of YouTube in 2005, there were few simple methods available for ordinary computer users who wanted to post videos online. With its easy-to-use interface, YouTube made it possible for anyone who could use a computer to post a video that millions of people could watch within a few minutes. It's hard to find someone who hasn't viewed a video on YouTube. From funny to educational, there are a huge number of videos available for viewing on YouTube. A video represents a great opportunity to create a personal connection with prospects and clients.

- SEO: It is the science and art of getting your company listed on the first search results pages. According to Sharpe (2008) "the Internet community is looking for information relevant to what potential customers are looking for, well-established companies that will provide the products and services that meet their customers' requirements. Search engines are looking for several qualities within each company's Website and beyond, and one major factor is the length of time a company is recognized on the Internet."

The internet marketing includes on-line advertising, banner ads, pop-up ads, superstitials, interstitials, micro sites, email and rich media banner ads (Fill, 2002). Also recently mobile marketing has been considerably developed in the form of ads and applications (e.g. iPhone applications).

In the field of tourism, hotels see the Internet as an opportunity to cost-effectively, (1) reach out to new customers via cyberspace and (2) enhance business efficiency among business partners. Some of the most common reasons for a hotel to go online may include taking reservations, promoting hotel operations, selling products and services, creating and establishing an identity or brand awareness, providing customer service and product support, generating repeat traffic, and advertising an event, product, or service (Sweeney, 2000).

As the Vorlow study (ibid: 29) notes, the Greek hotel enterprises that use the Internet for advertising their services have higher utilization of their room capacity at 4.8% and higher income at 3.7%. Consequently a study of the factors that influence the successful adoption of internet marketing activities would be particularly relevant.

1.3 Factors That Influence The Successful Adoption Of Internet Marketing Activities

There is limited published quantitative research on the internet marketing that has mainly taken the form of the use of the internet by consumers. This study differs in the sense that an attempt is made to identify and empirically verify the determinants of successful Internet marketing for Greek small tourism businesses. The factors (independent variables) considered in this study are those that are associated with the 'entrepreneur-manager issues' of the use of the Internet as a business/marketing tools relating to the perceived performance of the latter one.

The entrepreneur-manager -related factors are identified and empirically evaluated in this study through fourteen in-depth interviews with entrepreneurs-managers of tourist enterprises and specifically of hotels. These are the:

(a) personal involvement of the entrepreneur

(b) entrepreneur' s ability to select the suitable internet agency

(c) entrepreneur's ability to brief satisfactory the internet agency

(d) entrepreneur's understanding of web 2.0

(e) entrepreneur's writing skills for the web

(f) entrepreneur's marketing orientation.

1.4 Entrepreneur's involvement

Elliot & Boshoff (2007) states that the entrepreneur's involvement in the marketing activities of small tourist firms is directly, positively related to the perceived success of internet marketing. This study argues that delegating the day-to-day management of the small tourism businesses' Internet marketing to someone else, either inside or outside the business, is not considered satisfactory for the owner-manager, as this may compromise the success of Internet marketing. The explanation of that relates to the fact, that in small firms only the owner has usually the necessary knowledge to plan a successful internet marketing activity. By the term 'knowledge' the reference is made to the adequate understanding to their business, its products and services, the market, the competition, the pricing policies and the objectives that are to be achieved.

Usually, due to the small size of the firm, no written context exists about the aforementioned knowledge. The approach has a personal characteristic and includes everything that is in the entrepreneur's mind, and thus it is not easy for another person, either inside or outside the company, to have the 'whole picture'.

Blackburn & McClure (1998) argue that the entrepreneurs of the small firms do not possess the same abilities. Based on that, they have identified three types of them:

(a) enthusiasts-high IT skills, positive attitude, IT management focus

(b) pragmatists – low IT skills, pragmatic attitude, IT management focus

(c) artisans – low IT skills, unconvinced attitude, IT operations / administration focus.

Having that in mind, it can be assumed that the impact of entrepreneurs-managers involvement in the perceived performance of the marketing activities in the small firms will not be the same and will differ according to their abilities. In any case, Blackburn & McClure (ibid) have identified that the role of the entrepreneur-manager is central to the small firm innovation, new technology adoption and use.

1.5 Selection of the internet agency

According to Barnett (1997), 'while choosing an agency is rarely easy, it is often made more difficult than it needs because of the personalities of the people involved and their sometimes contradicting requirements'. Fill (2002) supports that the process for selecting an agency is relatively straightforward. To start with, a search is undertaken to develop a list of potential candidates. This step is accomplished by referring to the relevant publications resulting in a list where six or seven of them are expected to be included. Next, the client will visit each of the short listed agencies in what is referred to as a credential presentation. This is a crucial stage as it is now that the agency is evaluated for its degree of fitting with the client's expectations and requirements. The agency's track record, resources, areas of expertise and experience can all be made available on the Internet, from which it should be possible to short-list three or four agencies for the next stage in the process: "the pitch". To be able to make a suitable bid the agencies are given a brief and then required to make a formal presentation ("the pitch") to the client some 6-8 weeks later. By this presentation we can evaluate the way and the philosophy the agency approaches the strategic and creative issues of our interest. Finally, the project is awarded to the candidate who produces the most suitable proposal. The immediate selection process is finalized when terms and conditions are agreed and the winner is announced to the contestants and made public.

Figure 3: Steps when selecting the internet agency
(Fill, 2002)

1. Search to develop a list of potential candidates
2. Visits to each candidate-credential presentation
3. Further information about the candidates from the net
4. Brief and formal presentations
5. Final selection of the internet agency
6. Terms and conditions agreed

This formalized process is now being questioned as to its suitability. "Suitability" is a relative term, and a range of factors need to be considered. In fact, the selection process is a bringing together of two organizations who may have different expectations. However, their cooperative behavior is essential for these expectations to have any chance of materializing.

The first reason under question relates to the time of six to eight weeks. Today most companies need to find communication solutions in one than rather eight weeks. The second one relates to the internet agencies who feel that they have to invest a great deal into a pitch with little or no reward if the pitch failed. Their response has been to ask for payments to pitch which has not been received well by clients. The third reason refers to the tension that arises when each agency is required to generate creative ideas over which they have little control once a pitch has been lost. As a solution, companies could invite agencies to discuss mini-briefs (Jardine, 2000).

These discussion topics are considered to be essential and focused on related issues rather than being the traditional challenge about how to improve a brand's performance. By issuing a mini-brief on the day it saves the companies time and money (eliminates weeks of preparation and associated staff costs), while enables the client to see agency teams working together.

1.6 Briefing of the internet agency

Triki et al. (2007) point out that the qualitative briefing is an important role that a firm has to play in order to have a successful firm-agency relationship. The agency's work depends heavily on the quality and the quantity of the information provided by the firm's brief. Instead of following up the creation process, the agencies think that firms should concentrate on communication objectives so as to help the agency implement the advertising campaign. According to this research the clarity of briefing is the second most important factor for the successful outcome of the firm-agency relationship.

Hill (2006) relates briefing with the satisfaction of the agency's customers. In order to have a sustainable relationship with an agency, it is important to incorporate multiple measures of what consist a success during the briefing process. Fill (2002: 387) supports that "once an account has been signed a client brief is prepared which provides information about the client organization. It sets out the nature of the industry it operates in together with data about trends, market shares, customers, competitors and the problem that the agency is required to address".

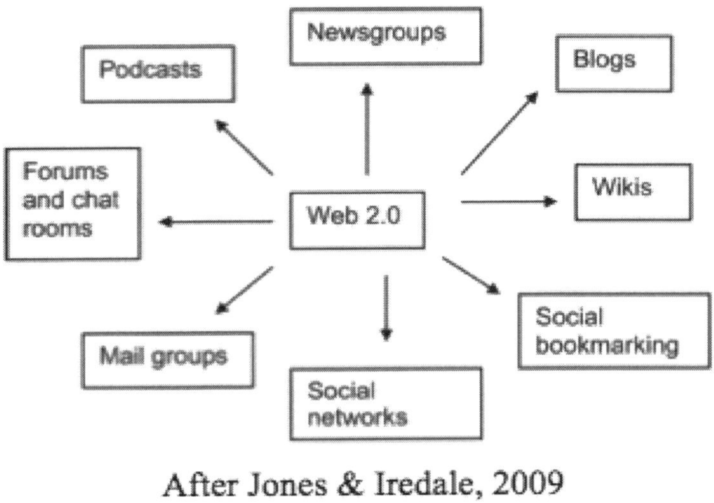

Figure 3. Components of Web 2.0

Briefing is used so that agency personnel to be well informed. In order the client problem is dealt with, the account planner will undertake research to determine a number of crucial fields such as: (i) market, (ii) media and audience characteristics. By determining that, proposals could be made to the rest of the account team as to how the client problem is to be resolved. Much of this information is translated into creative work for the development of copy and visuals by the appointed creative team. The briefing process provides the mechanism for the agency operation.

1.7 Understanding the Web 2.0

Web 2.0 refers to the second generation of web development and design. It is characterized as facilitating communication, information sharing, user-centered design and collaboration, and has fostered web based communities, hosted services, and web applications. "The essence of the social web is communication. The nature of the communication is continuously changed by technology as it becomes more interactive, democratic, participatory, immediate and responsive" (Jones & Iredale, 2009). Businesses have to recognize the need to work with their community of stakeholders as their image, brand, reputation and ultimately the bottom line can be negatively impacted or positively rewarded in the new online Web 2.0 world (Gelles, 2009).

According to Jones & Iredale (ibid) the Web 2.0 includes various tools such as: (i) social networks, (ii) media sharing and (iii) other interactive online tools. People who share similar interests can communicate or discuss, using those sites offered with similar orientation. *As far as the entrepreneurship is concerned, education, teaching, learning and assessment can find scope and space with Web 2.0.* Flexible online communities can be easily created, as well as an evolution which affects the range of the conventional media impacts.

Using the web, each person, at every level of the social hierarchy, has the opportunity to become the creator of a news story, than be a pathetic "consumer of news", challenging thus "the mainstream media's view of the world". In this sense, the web possesses all the characteristics of the most democratic tool of our times. Today, social networking has become the prevailing phenomenon and its growth has been tremendous as big numbers of the world population, business, state and private organizations, consumers and various interest groups "are increasingly engaging and involving themselves in new online communities of interest" (Jones & Iredale, ibid). YouTube is often mentioned as a market leader in this field.

One of the main differences that Web 2.0 has, in relation to Web 1.0, is that in the context of the former one branding and reputation are open to question, compliment and criticism. Shaping the brand image is "a joint process that requires the active involvement and willing consent of stakeholders, who now have a greater opportunity to make known their views" (Jones & Iredale, ibid). Introducing image, brad and reputation is a process that cannot be imposed from above, from the management side, since the consumers, the citizens and the suppliers are those who have the 'final word'. As the building of a relationship is necessary among the two parts of the market, Web 2.0 can be a useful tool in that direction. As Jones & Iredale (ibid) state, "Shared ownership of brand, information, image and reputation are key features of the social web environment".

Another interesting aspect of Web 2.0 is learning by monitoring other enterprises. Using the social web, a number of skills, behaviors and attributes can be enhanced. Also, it helps the individual to be in touch with the flexible market economy and society, to find employment, to start a business, to participate in a social activity. As a new development, the implications of Web 2.0 "are that businesses, consumers and other stakeholders are more informed and through the acquisition of knowledge they are empowered in new ways" (Jones & Iredale, ibid).

However, no matter how democratic a tool is, some prerequisites are necessary in order for the Web to be operational by customers, citizens or business managers. These prerequisites have to do with the appropriate knowledge of rights, needs and obligations. As citizens and activists enter the field of being active parts of the Web reality, play a part in shaping the new Web 2.0 landscape and this is something businesses should be aware of, alert to and prepared for. The phrase social-casting, as opposed broadcasting, captures the essence of the move brought about by the social web (Jones & Iredale, ibid).

1.8 Writing skills

The Wall Street Journal notes that eighty percent of businesses surveyed believe that their employees' biggest problem is written communication (Price, 2007). According to Wise (2005), writing for the web is a different style than other types of writing such as standard press releases. It needs to be shorter, more concise, more conversational, more enticing.

The way of writing for the web needs to comply with the way people read information on the Web. And this way is different than the one people read information in a newspaper or magazine. The visitor of a website has certain characteristics which are related to the specific of the information he is looking for. As a consequence, writing has to fulfill that prerequisite so that people can find information quickly. If the writing is not written for on-line, the reader will be probably lost. Moreover the on-line writing has to be tailored to the visuals on the page. A title and a few lines that will grab the reader's attention are necessary in order to attract the attention of the visitors; otherwise they will just keep browsing the internet. The website visitor's attention have to be captured immediately, it has to be an immediate connection.

Klipstine (2008) suggests five principles for improving online marketing writing: (1) Reading text from a computer screen is different from reading text from paper. People read 25 percent slower from a computer screen. (2) Be succinct. Write no more than 50 percent of the text you would write for a hardcopy publication. (3) Write for scanability. Users don't read word for word from a computer screen. And they don't like long blocks of text. (4) Use hypertext, headings, highlights and bulleted lists to break the material into chunks of information. (5) Writing for the Web should maintain the standards of solid journalism; writing should be mechanically excellent.

As far as the Greek internet agencies, they usually do not have adequate expertise in writing for websites. To deal with that, they either have to employ a skilled person or to address to a freelance. In fact, the Greek small tourist businesses do not usually have employees with writing skills. Thus, they have to use a freelance service. However, if the entrepreneur-manager would like to outsource this task to a freelance copywriter, he has to overcome two barriers for a successful collaboration. Firstly the cost of a freelance copywriter may be too high for a small firm. Secondly the entrepreneur-manager has to brief the copywriter so that the latter one undertakes the writing for the website or the online marketing activity in a suitable and satisfactory way. But, as has already been mentioned before, the adequate briefing is not always taken for granted.

1.9 Marketing orientation

Harding (1998: 35) states that the marketing orientation concept refers to "a set of organizational behaviours and techniques which develop value-added products derived from well-developed market research and consultation processes with loyal, long-term customers". Working in a highly competitive environment, organizations need to be very careful about the way they handle the customer. A company that understands the needs of its customer, and tries hard to develop and provide high-valued products, and market them as a whole across all departments, is considered a market-oriented company. In doing so, it embraces the 'marketing concept' that brings the customer into the center of its attention. The slogan used in this case is "the customer is King". The marketing concept is actually an attitude or a philosophy that characterises the whole management structure of a company.

Singh & Ranchhod (2004) state that there were four latent dimensions underlying this orientation and these are: customer orientation, competitor orientation, departmental responsiveness, and customer satisfaction orientation. From all four dimensions, customer orientation and customer satisfaction orientation have a stronger impact on performance.

Tzokas et al. (2001) mention Jaworski & Kohli's (1993) empirical evidence which suggests that "the adoption of a marketing orientation has a positive effect on business performance and also impacts on employee commitment". And also Deshpande et al. (1993, Tsokas et al., ibid) suggest that "as marketing is an organizational not a departmental function and requires a particular culture in order to adopt such an approach, issues of organizational culture are integral to discussions of marketing orientation". Implementing the idea of 'culture' is not an easy task since "definitions of culture in the literature vary and include ideology, a coherent set of beliefs, shared core values, important understandings, or the 'collective programming of the human mind' (Hofstede, 1980, in Tzokas et al., ibid).

Osborne (1996), referring to the role of strategic values writes that these are the rationale for the viability of a business and link the firm to its environment. And all this "because strategic values describe what a business does to win in the marketplace and why, they enable an organization to focus on what is important to success and determine what is irrelevant or counterproductive". Varey & Lewis (2000: 113) consider these values to be reflected in and be a reflection of "the prevailing culture within the organization".

Between the academics, there is a tendency "to adopt an interpretavist point of view, with culture being viewed as something an organization 'has', as compared with something an organization 'is ', or as the 'glue' which holds an the organization together" (Tzokas et al., ibid).

Diamantopoulos & Hart (1993: 103) copy Kohli & Jaworski (1990) arguing that "a market-oriented organization is one whose actions are consistent with the marketing concept". Having that in mind, we can argue that there develops "a positive linkage between the perception of the marketing concept as a business philosophy and the company actions underlying a market orientation" (Diamantopoulos & Hart, ibid).

The very fact that the product is co-produced and co-marketed with the consumer is the main reason why the application to the service sector has been modified (Li & Greenberg, 1997, in Harding, 1998). Harding (ibid: 39) goes on stating "Businesses with a marketing orientation adopt it because they want to exploit the desires of as many of the target market as the law and the competitors will allow –consumers are safe because they can always walk away". Thus, there has been an emphasis, within the marketing literature, on the importance of a firm attaining a marketing orientation in order to compete in today's marketplace.

2. METHODOLOGY

2.1 The measuring instrument

In order to ensure validity and reliability, measuring instruments from previous studies were used, where possible, to measure both the independent variables and the dependent variable. All the items in the questionnaire were linked to a 7-point Likert-type scale where 1 was labeled strongly agree and 7 was labeled strongly disagree in order to measure the dependent variable of the Perceived success of Internet marketing activities.

The sample

The sample is drawn by the population of Greek SMEs tourist businesses and specifically from those of the hotel sub-sector. According to the European Commission (2005), the classification of the European medium, small and micro companies (SMEs) is as follows in Table 1:

Table 1: Classification of the SMEs (European Commission, 2005)

Enterprise category	**Annual work unit**	**Annual turnover**	**Annual balance sheet total**
Medium sized	<250	<€ 50 million	<€ 43 million
Small	<50	<€ 10 million	<€ 10 million
Micro	<10	<€ 2 million	<€ 2 million

However, since companies in Greece that have not the Société Anonyme legal status, as it is usually the case with the small and micro enterprises, are not obliged to publish financial data, it is not possible to classify them according to the European Commission suggestions. For this reason, Papanikos (2000: 29) has developed a study which contains a combination of the SMEs classification according to the EC standards and the Hellenic Chamber of Hotels data. According to this study, the classification for the Greek hotels is as follows[3]:

Table 2: Classification for the Greek hotels (Papanikos, 2000)

				Employment	
Hotel size	**Number of rooms**	**Number of hotels**	**% of hotels**	**Average number of employees**	**Highest number of employees**
1. Family	1-20	3,548	43%	1,4	10
2. Small	21-50	3,011	37%	4,8	25
3. Medium	51-100	1.075	13%	15,0	60
4. Big	101 -	610	7%	64,0	210

It seems that the hotel size that corresponds to the SMEs definition is those of Family, Small and Medium ones. Consequently, for the purpose of the present research, the yardstick for selecting the SMEs hotels was that of the number of rooms, which, in the particular case, has been less than 101 rooms.

The sample was drawn from the Hellenic Chamber of Hotels database, where data for approximately 10.000 hotels in Greece can be found. A sample of 200 hotels that (i) have websites and (ii) have less than 101 rooms was randomly selected. These 200 hotels were surveyed the period October 2009 to December 2009 using a Web based questionnaire that allowed respondents to complete the questionnaire online. The usable 145 questionnaires received comfortably exceed the minimum guideline for the use of multivariate statistical procedures to analyze the data (Hair et al., 1998).

2.2 Statistical procedure

2.2.1 Validity assessment

The initial step in the data analysis was to assess the discriminant validity of the instrument used to measure the variables in the theoretical model. This was done by using the multivariate technique of exploratory factor analysis with the purpose of establishing which of the questionnaire items measure each construct in the theoretical model. The analysis was conducted by using the SPSS 13.0. In Table 3, it is displayed the factor structure that has been emerged by an iterative process, where items were deleted. Those deleted items either did not load higher than 0.45 on any factor or alternatively loaded higher than 0.45 or higher on two or more factors (cross-loading).

2.3 Reliability of the measuring instrument

Reliability is the consistency or stability of empirical indicators from measurement to measurement (Parasuraman, 1991). For measuring reliability the Cronbach's alpha coefficient is used, which is based on the average correlation of items within an instrument or scale and is regarded as an indication of internal consistency.

3. RESULTS

3.1 Discriminant validity results

The factor matrix of the exploratory factor analysis is shown in Table 1. It reveals that 27 items loaded on seven distinct factors and there were a total of 49.4 percent of the variance in the data. The seven factors were named Entepreneur Involvement (items EI01-EI04), Internet Agency Selection (items IS01-IS03), Internet Agency Briefing (items IB01-IB03), Understanding of the Web 2.0 (items UW01-UW03), Writing Skills (items WS01-WS03), Marketing Orientation (items MO01-MO04), Perceived Success of Online Marketing (items PM01-PM07). The items that measure each factor are described in Appendix 1.

3.2 Reliability results

The internal reliability of the seven factors that emerged from the exploratory factor analysis is shown in Table 4. All seven factors returned a Cronbach alpha coefficient score in excess of 0.70 and are accordingly considered reliable.

The hypotheses

The following hypotheses were empirically tested:

H1: There is a positive relationship between the entrepreneur's involvement in the internet marketing activities of their hotel and the perceived success of online marketing.

Table 3: Exploratory factor analysis

Item	Factor one Entepreneur's Involvement	Factor two Internet Agency Selection	Factor three Internet Agency Briefing	Factor four Understanding Web 2.0	Factor five Writing Skills	Factor six Marketing Orientation	Factor seven Perceived Success of Online Marketing
EI01	0,885						
EI02	0,823						
EI04	0,765						
EI03	0,654						
IS02		0,832					
IS02		0,689					
IS01		0,578					
IB01			0,743				
IB02			0,656				
IB03			0,576				
UW03				0,678			
UW02				0,574			
UW01				0,498			
WS01					0,862		
WS03					0,754		
WS02					0,672		
MO01						0,812	
MO03						0,796	
MO02						0,686	
MO04						0,578	
PM06							0,765
PM02							0,743
PM04							0,715
PM03							0,654
PM05							0,523
PM01							0,517
PM07							0,497

Table 4. Cronbach alpha coefficients

Factor	Cronbanch alpha
Factor one: Entepreneur's involvement	0,876
Factor two: Internet Agency Selection	0,734
Factor three: Internet Agency Briefing	0,798
Factor four: Understanding Web 2.0	0,834
Factor five: Writing Skills	0,745
Factor six: Marketing Orientation	0,723
Factor seven: Perceived Success of Online Marketing	0,854

H2: There is a positive relationship between the entrepreneur's ability to select the suitable internet agency and the perceived success of online marketing.

H3: There is a positive relationship between the entrepreneur's ability to brief adequately the internet agency and the perceived success of online marketing.

H4: There is a positive relationship between the entrepreneur's understanding of the Web 2.0 and the perceived success of online marketing.

H5: There is a positive relationship between the entrepreneur's writing skills and the perceived success of online marketing.

H6: There is a positive relationship between the entrepreneur's marketing orientation and the perceived success of online marketing.

3.3 Multiple regression analysis results

A multiple regression analysis has been conducted to address the aforementioned six hypotheses. The results, with the five independent variables – entrepreneur's involvement, internet agency selection, internet agency briefing, understanding of web 2.0, writing skills, marketing orientation – and the perceived value on online marketing, as the dependent one, are shown in Table 5.

It is obvious from the Table 5 that the aforementioned independent variables exert a positive influence on the dependent variable; consequently the hypotheses H1, H2, H3, H4, H5, H6 are accepted.

Table 5: Multiple regressions results

Source	df	SS	Mean Square	F	Pr > F
Model	6	196,912	32,81	60,231	0,0001
Error	311	200,563	0,645		
Corected total	317	397,475			

$R^2 = 49{,}5\%$

Parameter	Estimate	t-values	Exceedance	Std error
PERSUCINTER	2,754	0,005	0,334	
E_M involvement	0,308	4,834	0,001	0,050
I_A selection	0,153	2,943	0,005	0,054
I_A briefing	0,191	2,832	0,005	0,052
U_W 2.0	0,165	3,187	0,003	0,051
W_Skills	0,145	2,814	0,005	0,054
M_Orientation	0,121	2,341	0,030	0,056

3.4 Managerial Implications

A hotel should use all modern means to promote itself and be known at the public (Karagiannis, 2001). It is widely accepted that the Internet is a medium that allows a targeted and cost-effective way to approach target groups either locally or internationally, especially the latter one. Since the SM Greek hotels can not afford the luxury of using the traditional media, due to budget restraints, the Internet seems to be the ideal media for achieving their objectives. The entrepreneur-manager of a SM Greek hotel is the most suitable person to be involved in the online marketing of his enterprise due to his overall understanding of the business, his high motivation to achieve the corporate objectives and due to the luck of specialized employees.

In order to fully take advantage of the Internet potentialities, the entrepreneur-manager has to be personally involved in the planning and implementation of the online marketing activities. In doing so, he will obtain:

(i) The appropriate know-how to select and brief the Internet agency to collaborate with,

(ii) An adequate understanding of the Web 2.0,

(iii)The adequate writing skills for the web, and

(in) A marketing orientation

Since it is difficult for all these qualities to be found in an SM Greek hotel's entrepreneur-manager, it is highly recommended the development of suitable courses that will address the aforementioned issues. These courses have to be quite practical and focused on "how to" rather on "why" so as to be immediately utilized. Also the SM Greek hotel's entrepreneur-managers are strongly advised to fully use the Internet for personal or professional purposes (e.g. social networking sites, professional sites, email, blogging and mirco-blogging, chatting, cloud computing, publishing electronic material etc.) so as to have an experience at first hand about the potentialities of the latter one as well as to keep pace with its continuous developments.

In any case, since the matter presents specific interest, a deeper research would be more useful. More objective assessments of marketing success and the inclusion of a wider variety of variables that can drive Internet marketing success leave scope for future research.

REFERENCES

1. Anderson S. (2009), 'Diving into internet marketing', American Agent & Broker, December.
2. Blackburn, R. and R. McClure (1998), The Use of Information and Communication Technologies (ICTs) in Small Business Service Sector Firms, London: Small Business Research Centre, Kingston Business School.
3. Barnett, P.(1997), Nike dumps UK shop, Marketing, 27 November, p.15.
4. Buhalis D. <ed> (2001), Tourism distribution channels, Practices, issues & transformations, Continuum, London.
5. Cooper Ch., Fletcher J., Gilbert D., Wanhill S. (1998), Tourism, principles and practice, Prentice Hall-Financial Times, Harlow (2nd edition).
6. Diamantopoulos A., Hart Susan (1993) Linking market orientation and company performance: preliminary evidence on Kohli and Jaworski's framework, Journal of Strategic Marketing, 1 (2), pp. 93-121.
7. Elliott, R., Boshoff, C. (2007), The influence of the owner-manager of small tourism businesses on the success of internet marketing, S. African Journal of Business Management, 38 (3).
8. European Commission (2005), The new SME definition, Enterprise and industry publications, pp. 14.
9. Fill, C. (2002), Marketing Communications, Prentice Hall-Financial Times, (3rd edition)
10. Gelles, D. (2009), Blogs that spin a web of deception, Financial Times, Vol. 12, February, pp.14.
11. Gibb, A.A. (1993), The enterprise culture and education: understanding enterprise education and its links with small business, entrepreneurship and wider entrepreneurial goals, International Small Business Journal, 11 (3), pp.11-34.

12. Gilmore, A., Carson, D. & Grant, K. (2001), SME marketing in practice, Market Intelligence and Planning, 19 (1), pp. 6-11.

13. Hair, J. F., Anderson, R. E., Tatham, R. L. & Black, W. C. (1998), Multivariate data analysis, 5th Edition. Englewood Cliffs, New Jersey: Prentice Hall.

14. Harding S. (1998) The marketing orientation and nonprofits: the concept revisited, Social Marketing Quarterly, 4 (4), pp. 35-39.

15. Hill,R. (2006), Advertiser satisfaction with advertising agency creative product, European Journal of Marketing , 40 (11/12), pp. 1254-1270.

16. Jardine, A. (2000), Will workshops replace the pitch?, Marketing, 13 April, p.16.

17. Jones, B., Iredale, Norma, (2009), Entrepreneurship education and Web 2.0, Journal of Research in Marketing and Entrepreneurship, 11 (1), pp. 66-77.

18. Karagiannis St. (2004) Productivity and Managerial efficiency in Hotel Industry. Crete as a case study, Tourist Scientific Review-Semestrial Journal, V. 2, pp. 167-181.

19. Karagiannis St. (2001) The prospectives of hotel marketing and the contribution of Hotel manager. The case of Crete, Dioikitiki Enimerosi, Vol. 21, pp. 104-114 (in Greek).

20. Klipstine,T. (2008), Time to take our own advice: Five simple principles for effective Web writing, Tactics, February

21. Lovelock C., Wirtz J. (2007), Services Marketing, Pearson International Edition, p.16-21.

22. Marcati A., Guido G. & Peluso A. (2008), The role of SME entrepreneurs' innovativeness and personality in the adoption of innovations, Research Policy, 37 (9), pp. 1579-1590.

23. Middleton V. (1993), Marketing in Travel & Tourism, Butterworth Heinemann, Oxford.

24. Osborne, R. (1996), Strategic values: the corporate performance engine, Business Horizons, 39, September/October, pp. 41–48.

25. Pavlopoulos P. (2007) Tourism establishment small and medium size: role, perspectives, measures, Institute of Tourism research and Predictions (ITEΠ), Athens (in Greek).

26. Papanikos G. Γ. (2000), The Greek SM Hotel Enterprises, Institute of Tourism research and Predictions (ITEΠ), Athens (in Greek).

27. Parasuraman, A. (1991), Marketing research, 2nd Edition,Reading: Addison-Wesley.

28. Price,H., (2007), Poor Writing Skills Cost You Money, South Carolina Business, May.

29. Sampaniotis Th., Vorlow K. (2006) Tourism & Greek Economy, Economy & Markets magazine, Eurobank Research, Vol. 8/ Year I, October (in Greek).

30. Sharpe J. (2008), What to know about Internet Marketing, Air-conditioning, Heating & Refrigerated News, October.

31. Singh S., Ranchhod A. (2004) Market orientation and customer satisfaction: Evidence from British machine tool industry, Industrial Marketing Management, 33, pp. 135– 144.

32. Sweeney, S. (2000), Internet marketing for your tourism business, Gulf Breeze, FL: Maximum Press.

33. Triki,A., Redjeb,N., Kamoun,I. (2007), Exploring the determinants of success/failure of the advertising agency-firm relationship, Qualitative Market Research: An International Journal, 10 (1), pp. 10-27.

34. Tzokas N., Carter Sara, Kyriazopoulos P. (2001) Marketing and Entrepreneurial Orientation in Small Firms, Enterprise and Innovation Management Studies, 2 (1), pp. 19–33.

35. Varey R., Lewis Barbara (2000) Internal marketing: directions for management, Routledge, London.

36. Vorlow K. (2008) Internet and the Hotel sector in Greece, Economy & Markets magazine, Eurobank Research, 3 (1) (in Greek).

37. Voulgaris, F., Asteriou, D., & Agiomirgianakis, C. (2004), Size and Determinants of Capital Structure in the Greek Manufacturing Structure, International Review of Applied Economics, 18 (2), 247-262.

38. Wise, K. (2005), The importance of writing skills, Public Relations Quarterly, 50 (2), pp. 37-40.

39. Zimmerer, T.W. & Scarborough, N.M. (2002), Essentials of entrepreneurship and small business management, 3rd Edition, New Jersey: Prentice Hall.

APPENDIX 1

EI01	The entepreneur-manager is the main participant in the planning of their hotel's on-line marketing activities
EI02	The entepreneur-manager is the main participant in the implementation of their hotel's on-line marketing activities
EI03	The entepreneur-manager consistently monitors the results of their hote's on-line marketing activities
EI04	The entepreneur-manager takes corrective action in case the results of their hotel's on-line marketing activities are not the desired ones
IS01	The entepreneur-manager is aware about the process for selecting the internet agency
IS02	The entepreneur-manager knows what to look for in selecting the internet agency
IS03	The entepreneur-manager is experienced in selecting an internet agency
IB01	The entepreneur-manager knows how to brief the internet agency about their hotel
IB02	The entepreneur-manager knows how to brief the internet agency about what they would like to achieve via the online marketing activities
IB03	The entepreneur-manager knows how to set objectives for the online marketing activities
UW01	The entepreneur-manager uses social media
UW02	The entepreneur-manager participates to bloggs
UW03	The entepreneur-manager uses the Internet to communicate and share electronic material with customers, friends, employees
WS01	The entepreneur-manager is able to be succinct in their writing
WS02	The entepreneur-manager is able to break-down the information into understandable and attractive chuncks of information
WS03	The entepreneur-manager is able to motivote via their writing the propsect to ask for more information or book a room
MO01	The entepreneur-manager is focused on their customer satisfaction
MO02	The entepreneur-manager monitors and takes into consideration their competitors activities
MO03	The entepreneur-manager is focused on profitability
MO04	The entepreneur-manager integrates their marketing activities
PM01	The online marketing activities results in more rooms booking
PM02	The online marketing activities results in increased customer satisfaction
PM03	The online marketing activities results in enhancement of hotel's brand
PM04	The online marketing activities results in increased customer refferals
PM05	The online marketing activities results in increased customer service
PM06	The onlive marketing activities are a value-for-money promotional tool
PM07	The online marketing activities increase hotel's customer loyalty

Chapter 3

An Assessment of the Relationship between Managerial Competencies and Exemplary Leadership Practices: Case of the Lodging Industry in Lebanon

Maraouch Fadila

Department of Hospitality Management, American University of Science and Technology, Beirut-Lebanon

ABSTRACT

Facing the rising demand for improved quality, reduced costs, and constant innovation necessitates to continuously up-grade knowledge, skills, and abilities as essential for all leadership roles including those held by top management. With management scholars racing to meet the aforementioned demand, "Competency Modeling" has become one of the most frequently used techniques to help organizations profile jobs, select, as well as prepare the right person for the right leadership position. Moreover, scholars introduced the "Exemplary Leadership" model which aims at helping managers lead their organizations more effectively where there is a belief that the "great person" theory is not accurate since anyone can aspire to become an exemplary leader by following the five practices characterizing the former model.

In general, the lodging industry has plenty to benefit from adopting competency models in identifying competencies necessary for current and aspiring leaders of lodging establishments and to stay competitive it must identify the competencies required for leadership not only for the present but also for the future.

The aim of this explanatory research is to examine the relationship among Managerial Competencies, and Exemplary Leadership practices of managers in the Lebanese lodging industry. The design was a non-experimental, quantitative, explanatory, correlational design, and has been conducted through an onsite survey to collect data. The survey used a modified version adopted with permission from of Koenigsfeld'sManagerial Competency, and the Kouzes and Posner's LPI surveys. The target population consisted of 254 conveniently selected general and department managers of hotels in Lebanon. The methods of data analysis included descriptive statistics, Pearson product moment correlation, factor analysis, and hierarchical (forward) regression analyses that tested the research questions. Also, scales were examined for reliability and validity.

Findings indicated that there exists a relationship among the Managerial Competencies (Conceptual/Creative, Leadership, and Interpersonal); and Exemplary Leadership practices of managers in the Lebanese lodging industry. Outcomes of the research help explore other industries in the country where different business environments exist as well as replication in other countries to further extend the body of knowledge of Leadership practices globally.

KEYWORDS

Competencies, Exemplary, Leadership, Lebanon

1 INTRODUCTION

In facing the rising demand for improved quality, reduced costs, and constant innovation, the need to continuously upgrade knowledge, skills, and abilities is essential for all leadership roles including those held by executives, managers, and supervisors (Kay & Russette, 2000; Mirabile, 1997; Okeiyi, Finley, & Postel, 1994; Tas, LaBreque, & Clayton, 1996). With management scholars racing to meet this demand, *competency modeling* has become one of the most frequently used techniques to help organizations profile jobs, select, as well as prepare the right person for the right leadership position (Katz, 1955; Sandwith, 1993).

Under the auspices of transformational leadership, Kouzes and Posner (1995) introduced the *Exemplary Leadership* model which aims at helping managers lead their organizations more effectively. The authors believe that the "great person" theory is not accurate since anyone can aspire to become an exemplary leader by following the five practices in their model (Goldsmith, 2007).

The purpose of this explanatory (correlational) study was to examine the relationship among the *Managerial Competencies* and *Exemplary Leadership* practices of managers in the Lebanese lodging industry. The design was a quantitative, explanatory, correlational design, and has been conducted through an onsite survey to collect data. The target population consisted of 252 general managers and department managers of hotels in Lebanon with a sample design that was considered a convenience sampling plan. The methods of data analysis included descriptive statistics, Pearson product moment correlation, factor analysis, and hierarchical (forward) regression analyses that tested the research questions. Finally, all scales were examined for reliability and validity.

2 LITERATURE REVIEW

2.1 Leadership

The *New York Times* has rated Beirut as the number one destination to visit among forty-four most compelling destinations around the world (Sherwood & Williams, 2009). Lebanon has been in-and-out of the travel and tourism spotlight since its existence. Internal and external civil unrest and wars have affected the country on all frontiers: socially, financially, and politically. However, relative stability in the last two decades is finally starting to pay off especially in the travel and tourism sector which constitutes a major part of the country's economy (Lebanon, 2010).

Leadership is defined as "the ability to inspire confidence and support among the people who are needed to achieve organizational goals" (Dubrin, 2007, p.2). It is understood by many to imply "collective action, orchestrated in such a way as to bring about significant change while raising the competencies and motivation of all those involved-that is, action where more than one individual influences the process" (Bornstein & Smith, 1996, p. 282). In this study, the focus was on one specific type of leadership, *Exemplary Leadership* as theorized by Kouzes and Posner (1995).

For the past three decades, the most dominant leadership theory is concerned with differentiating between two leadership styles: *transactional*and *transformational* (Tourangeau & McGilton, 2004). Simply stated, transactional leadership is based on the concept of an exchange between leaders and followers while transformational leadership goes beyond such an exchange to a level where the leader uses personal traits and relationships to influence followers (Nahavandi, 2009). Based on such interaction, a transformation of attitudes, values, and behaviors occurs which facilitates major organizational change as well as increases organizational performance (Burns, 1978). Consequently, organizations that are anticipating or embarking on major organizational changes would greatly benefit from implementing transformational leadership practices. To achieve this, organizations are realizing that leadership is needed at all levels of the organization and not necessarily exclusively found among people in high levels of the organization. Although such a statement can be promising for aspiring leaders, this is easier said than done since developing leadership skills is a much more complex process than developing a structured skill (Dubrin, 2007).

Furthermore, scholarly research asserts the need to constantly develop leadership theories in order to maintain not only competitiveness but also sustainability. *Exemplary Leadership* as a form of transformational leadership is among the latest leadership theories that is achieving promising results (Kouzes & Posner, 1995). Proclaiming that leadership is not a reserve for a few lucky men and women, Kouzes and Posner (1995) identified five *Exemplary Leadership* practices that enable leaders to get extraordinary things done. Consequently, studying the relationship between*Managerial Competencies* and *Exemplary Leadership* is a new area of study associated with leadership effectiveness.

2.2 Exemplary Leadership

Exemplary Leadership is a theoretical framework developed by Kouzes and Posner that has its roots in transformational leadership (1995). Based on their research, Kouzes and Posner (1995) identified human relations skills as the means through which leaders promote success within the organization. After studying thousands of best leadership experiences, where leaders performed at their best, the authors identified five *Exemplary Leadership* practices common to successful and effective leadership that can be adopted by anyone who accepts the leadership challenge. These are: *"model the way, inspire a shared vision, challenge the process, enable others to act,* and *encourage the heart"* (Kouzes and Posner, 1995). Each practice includes two commitments that were identified as leading to exemplary practice.

Kouzes and Posner measured *Exemplary Leadership* using the Leadership Practice Inventory (LPI). The LPI is made up of thirty questions from which every six assess one practice. These questions describe various leadership actions and behaviors reflected in the five practices mentioned earlier. There are two versions of the LPI-self and observer. The self-assessment is completed by the individual leader while the observer assessment is completed by a peer, superior, or subordinate of the leader (Kouzes and Posner, 1995). In a meta-analysis of 18 different leadership competencies, the LPI received top scores both in psychometric soundness and in ease of use (Huber et al., 2000).

2.3 Managerial Competencies

Understanding and studying managerial skills or competencies is a relatively new area of study that has started with the research of scholars such as Katz (1955) and McClelland (1973). Over the past twenty years, the question of what competencies managers need to be effective in their jobs has changed considerably from an emphasis on technical abilities in the 1970s and early 1980s to a more personnel-management abilities focus in the late 1980s until the present time (Katz, 1955; Sandwith, 1993; Tas et al., 1996).

A competency model is defined as the "output from analyses that differentiates high performers from average and low performers" (Mirabile, 1997, p. 75). Leadership competency models based on performance rather than traits have been the focus of many researchers in the second part of the last century (Chung-Herrera, Enz & Lankau, 2003; Katz, 1955; McClelland, 1973; Sandwith, 1993). This focus was due to the fact that performance is easier to identify and measure and is less likely to be misinterpreted. The motive was to present a more useful approach that aims at understanding what good executives *do* (i.e., skills exhibited in carrying out jobs effectively) instead of what good executives *are* (i.e., innate traits and characteristics) (Katz, 1955, p.33).

Katz's approach is based on firsthand observation of executives at work coupled with field research in administration and suggested that there are three basic developable skills which necessitate the need for identifying specific traits that can also provide a useful way evaluating the administrative process. These are: (1) *Technical,* (2) *Human,* and (3) *Conceptual* (Katz, 1955, p. 34). In comparing and contrasting the above skills, the relative importance of the *Technical* skill lies at the lower levels of the organization. As the administrator moves further from the actual physical operation, the need for *Technical* skill diminishes to the point it might become nonexistent due to the existence of the *Technical* skills of others who took the administrator's position. At this level, the administrator might still be able to perform effectively if the other two skills, *Human* and *Conceptual,*are adequately developed. Katz did not provide measurements for his constructs; therefore, reliability and validity were not discussed

Since the publication of Katz's (1955) study, many other researchers have used this typology of research. The most notable is the work of Sandwith (1993). Sandwith (1993) developed a model of competency *Domain*s that guides large organizations in their management training and development programs. Sandwith's (1993) model had its origin in the three skills approach that was initially addressed by Katz (1955) in his work *Skills of an Effective Administrator*. In the following years, other renowned researchers such as Mintzberg discussed this model and its application by a number of organizations (Sandwith, 1993). These researchers found that there exists a broader range of activities used by managers than was previously conceived. As a result, the *Human* skill factor of the original concept was becoming too general.

To solve this problem, Sandwith (1993) updated this category and divided it into three separate *Domain*s which he labeled: *Leadership, Interpersonal,* and *Administrative*. Consequently, Katz's (1955) original three category model was expanded to include five areas of *Managerial*

Competencies or *Domain*s (noting that a manager's work is rarely confined to one Domain). These included (p.46):

Conceptual/Creative Domain: this refers to the cognitive skills associated with comprehending important elements of the job. It should be noted that the creative term was added to encompass recent understanding of functions related to brain and creative thought.

The Leadership Domain: this domain provides a strategic link between the *Conceptual/Creative Domain* and all other domains.

The Interpersonal Domain: this domain focuses on the skills for effective interaction with others.

The Administrative Domain: this domain addresses the activities that lie between the *Interpersonal Domain* and the *Technical Domain*. It refers to the personnel management and financial management aspects and does not mean paper work and processes that are found in most aspects of the organization.

The Technical Domain: this refers to the actual work the organization does such as production standards; work processes; monitoring, reporting and evaluating processes; and the knowledge and skills required to implement them.

Sandwith (1993) claimed that a competency model based on these five *Domain*s can be very useful for organizations because: (1) the model can develop a competency profile for a particular job to identify the knowledge and skills requirements, (2) the competency requirements of an individual can be matched to that for a particular job, (3) appropriate learning tools can be identified to accommodate competency needs in one or more *Domain*s, and (4) the model can provide a complete set of managerial competency profiles which can aid in creating a comprehensive framework for training and development efforts of organizations (p.45).

The Managerial Competencies Model is socially significant addressing essential issues about hiring, training, and promoting techniques in the discipline of management in general and sociology and psychology in particular; and is useful in describing and predicting managerial effectiveness among those aspiring leadership positions. Later, various researchers such as Kay and Russette (2000) and Tas et al. (1996) conducted studies that verified the propositions, or five areas of *Managerial Competencies*, providing empirical validity to this model. The major proposition with conflicting results is related to the ranking of *Technical* over *Interpersonal* competencies which necessitates the need for further empirical evaluation in this area (Kay & Russette, 2000) while the most useful proposition is the dominance of leadership competencies in those same studies. The model has been adapted to different situations and populations with over seventy studies referencing its use in recent management history.

The idea of testing for competence in the hospitality industry is credited to Richard Tas (1988) who conducted what is considered the pioneering study that aimed at identifying important competencies for hotel general manager trainees. Tas (1988) introduced a conceptual framework that identified three categories: Essential Competencies, Considerably Important Competencies, and Moderately Important Competencies. Tas suggested that the most important competencies were focused on human relations skills.

Tas's (1988) conceptual framework has been used by many scholars in the field such as Baum (1990) and Okeiyi et al. (1994); and had been adapted by Tas et al. (1996) to include another competency model by Sandwith (1993). Several empirical studies such as that by Kay and Russette (2000) led to a special focus on Essential Competencies (EC) and their significance in selection and recruitment of managers as well as in the preparation of hospitality programs curricula.

Examples of major studies that followed Tas's suit are those belonging to Okeiyi et al. (1994) that focused on competencies in F&B management; Tas et al. (1996) that focused on competencies related to property management; Kay and Russette (2000) that focused on competencies related to general hospitality management; and Chung-Herrera et al. (2003) that presented a model focused on leadership competencies needed for future hospitality managers.

Managerial Competencies in the lodging industry were measured using the club manager competency survey developed by Koenigsfeld (2007). The original survey consisted of three sections: club and manager demographics, *Managerial Competencies*, and ability and success measure with a total of 356 questions. Of those, the second section consisted of 151 *Managerial Competencies* that were placed into five *Domain*s according to Sandwith's (1993) *Competency Domain Model*.

For the purposes of this paper, the *Managerial Competencies* developed by Koenigsfeld (2007) were adapted to reflect those of the lodging industry instead of the club industry. The same five-point Likert scale was used to measure the *Managerial Competencies* of managers in the Lebanese lodging industry. Worth mentioning that the competency model developed by Koenigsfeld (2007) was used with permission for adaptation to fit the purpose of the study. Moreover, lodging managers, in this study, are defined as those who are 18 years of age or older, speak and write English fluently, are currently employed at a hotel in Lebanon, and who are holding the title of at least a department manager in the hotel.

2.4 Research Justification

Today, most large international companies including hospitality companies, are spread over the world with leadership teams managing the global operations; such teams and their leaders have a critical task ahead of them due to "highly diverse cultural bases, little face-to-face contact, and infrequent full-scale team meetings. Their leaders are projected to them electronically or via videoconference. They must constantly deal with the issue of weak commitment, divergent values and ambiguous power bases" (Miller, 2001, p. 26-27). Evidently, disasters happen and whether they are natural or manmade, successful organizations are realizing the importance of leadership and its valuable role in steering the operation to safe harbor with the least possible damage.

Effective leadership and *Managerial Competencies* is a topic of local and global interest (Chapman & Lovell, 2006; Chung-Herrera et al., 2003; Katz, 1955; Sandwith, 1993). Effective and competent leaders have been identified as crucial to organizations wanting to thrive locally, nationally, or globally. From there, competency models have started to emerge based on studies from various researchers who preached their usefulness on multi-levels such as increasing organizational performance, reducing executive turnover rates, improving training and development programs, and improving educational programs to offer more compatible curricula. As such, having and continuously updating competency modeling can influence the performance of hospitality industry leaders in general and the lodging industry leaders in particular to show better performance in competitive times (Dalton, 1997; Kay & Russette, 2000; Mirabile, 1997; Okeiyi et al., 1994; Perdue et al., 2002; Sandwith, 1993; Tas, 1988; Tas et al., 1996).

It is noteworthy to mention that the investigation of *Managerial Competencies* in the lodging industry (before investigating relationships with other variables such as *Exemplary Leadership*) is a new area of study that has only recently attracted the attention of researchers and practitioners in the field (Chung-Herrera et al., 2003; Kay & Russette, 2000; Okeiyi et al., 1994; Perdue et al., 2002; Tas, 1988; Tas et al., 1996). While recommendations of studies done so far suggest the need to pursue research in more hospitality-related industries and in various geographical locations; preliminary research using databases of scholarly, peer-reviewed s yielded no results of *Managerial Competencies* and its relationship to *Exemplary Leadership* in the lodging industry. Moreover, no results were found of such relationship as pertaining to the country of Lebanon. This was a major gap in the literature that is hoped to be filled in by extending research in this field.

The topic of this study is researchable because all variables, research questions, and hypotheses could be measured by scientific questionnaires and statistical analysis. This study is also feasible, because it can be implemented under reasonable time and cost limits.

2.5 Research Question

The aim of this study was to determine if there is a relationship between *Managerial Competencies* and *Exemplary Leadership* practices. There is one major research question accompanied by one main hypothesis (H) and five (5) sub-hypotheses. The main research question of this study was to empirically investigate the relationship between Lebanese lodging industry managers' perceived importance of *Managerial Competencies* and*Exemplary Leadership* practices.

2.6 Research Hypotheses

H1: There is a significant positive relationship between perceived importance of Managerial Competencies and Exemplary Leadership practices of managers in the Lebanese lodging industry.

H1a: There is a significant positive relationship between perceived importance of *Managerial Competencies* and the *"model the way"* practice of managers in the Lebanese lodging industry.

H1b: There is a significant positive relationship between perceived importance of *Managerial Competencies* and the *"inspire a shared vision"* practice of managers in the Lebanese lodging industry.

H1c: There is a significant positive relationship between perceived importance of *Managerial Competencies* and the *"challenge the process"* practice of managers in the Lebanese lodging industry.

H1d: There is a significant positive relationship between perceived importance of *Managerial Competencies* and the *"enable others to act"* practice of managers in the Lebanese lodging industry.

H1e: There is a significant positive relationship between perceived importance of *Managerial Competencies* and the *"encourage the heart"* practice of managers in the Lebanese lodging industry.

3 METHODOLOGY

3.1 Research Design

To examine the relationship among the *Managerial Competencies* and *Exemplary Leadership* practices of managers in the lodging industry, a study was conducted on managers of lodging properties in Lebanon. The design was a non-experimental, quantitative, explanatory, correlational design, and was conducted through an onsite survey to collect data. An explanatory research design is a "correlational design in which the researcher is interested in the extent to which two variables (or more) co-vary, that is, where the change in one variable is reflected in changes in the other" (Creswell, 2005, p. 327). Based on the literature review, there are no empirical studies that have examined the research question and hypotheses outlined in this study.

The purpose of this research design was to test one hypothesis with H1 having five sub-hypotheses. Multiple regression analysis was utilized with SPSS to test the aforementioned hypotheses. The independent variables are the *Managerial Competencies* of managers in the

Lebanese lodging industry; the dependent variable is *Exemplary Leadership* practices.

In this study, lodging properties consisted of property characteristics that include sleeping accommodations among an array of services. *Managerial Competencies* referred to the *Conceptual/ Creative, Interpersonal, and Leadership* skills of managers based on a model developed by Sandwith (1993); and tested by Koenigsfeld (2007) in the club management industry. Finally, *Exemplary Leadership* referred to five practices identified by Kouzes and Posner (1995) (*model the way, inspire a shared vision, challenge the process, enable others to act,* and *encourage the heart*) as essential for success as an effective manager.

The survey contained three parts where Part A covered the *Demographic Profile* of Lebanese managers participating in the survey, Part B covered the*Managerial Competencies* adapted from the instrument developed by Koenigsfeld (2007), with permission, and part C covered the five practices of*Exemplary Leadership* developed by Kouzes and Posner (1995) with permission.

3.2 Sampling Plan

The target population consisted of general managers and department managers of all hotels in Lebanon. Due to the relatively small size of the country and in an effect to maximize responses, the accessible population was the same as the target population. The main entity contacted is the Ministry of Tourism since all hotels are required to obtain permission to operate and comply with the Ministry's rules and regulations. Also, to ensure anonymity, only hotels with a minimum of five managers will be contacted to participate in the survey. The selection of participants was based on certain criteria to enhance the internal validity of the study. For inclusion in this study, participants must be:

18 years or older

English speaking

Currently employed at a hotel in Lebanon

Holding the title of General manager or, at the minimum, a Department Manager of the hotel

The target and accessible population were the same for this study: the General Manager and Department Heads holding the title of at least a Manager in a hotel whose position is indicative of leadership to a group of hotel employees. Therefore, the final sample consisted of hotel managers and department managers who responded to the researcher's request for administering the survey on them and who fit the eligibility criteria for this research.

Because this plan is categorized as a convenience sampling plan, the participation rate is not a key figure to mention since only willing and eligible employees participated in the survey. Nevertheless, not all returned surveys were usable surveys.

In this study, a total of thirty hotels were contacted to participate by sending an email to the general manager and/or human resource manager of each hotel. From these, and over a period of two months, a total of 284 surveys were picked up from the various participating hotels resulting in 254 usable surveys which constitutes a valid response rate of 89% (254/284).

3.3 Survey Instruments

For this research, the questionnaire to be used was made up of three survey instruments. Part A was developed by the researcher and measured the*Demographic Profile* of managers/leaders, Part B measured the *Managerial Competencies* of managers/leaders using an adapted version of the Club Manager Competency Survey developed by Koenigsfeld (2007). Part C measured the five practices of *Exemplary Leadership* using the LPI instrument developed by Kouzes and Posner (1995).

3.4 Data Collection and Analysis Methods

Hotels with a minimum of five managers were contacted. Permission was requested from General Managers of hotels prior to delivering and administering surveys. Completion of the survey by respondents was anonymous. A box with a slit on top was left at the Front Desk where respondents dropped off their finished surveys.

The data collected from the survey was analyzed using the Statistical Product and Service Solutions, (SPSS) version 19.0, an IBM product acquired by IBM in 2009 (Hejase and Hejase, 2013, p. 57). The methods of data analysis included descriptive statistics, Pearson product moment correlation, factor analysis and multiple regression. Validity of the instrument was enhanced by having Parts A, B, and C reviewed by a panel of experts. All hypotheses were tested using a 5 percent level of significance:

To test H1, if there is a significant positive relationship between *Managerial Competencies* and *Exemplary Leadership* practices, multiple regression was used.

4 RESULTS AND FINDINGS

4.1 Demographics

The sample indicated that respondents were almost entirely Lebanese (94.1%) with North Americans (1.2%) and Western Europeans (2.8%) comprising less than five percent of the sample. The majority of respondents were between 18 and 34 years old (75.5%). Respondents over the age of 65 represented the smallest age group (1.2%). Almost two thirds of respondents were males (69.3%) and about one third was females (30.7%).

The majority of respondents had an undergraduate degree (44.1%) followed by those with a graduate degree (25.6%) while those with a doctoral degree represented the smallest percentage of respondents (2.4%). A little less than two third (60.6%) of respondents studied hospitality management during their education period.

The majority of respondents had an average of three years of professional experience in the lodging industry (40.9%) followed by those with an average of eight years of experience (29.9%) with few respondents having over twenty years of experience in the lodging industry (2.4%). Almost two thirds of respondents had an average of three years of management experience in the lodging industry (67.7%) followed by those who have an average of eight years of management experience in the lodging industry (19.7%). Most respondents had three years as an average for years employed in current position (83.9%).

The majority of respondents worked in hotel properties that had an average of 150 rooms (41.7%). Properties that had an average of 450 or more rooms represented the lowest category (1.6%). As far as number of employees per property, 54.3% worked in hotels that employed an average of 150 employees while only 3.6% worked in properties that employed an average of 450 or more employees.

Since only four and five star hotels were selected for the study, respondents who worked for five star hotels as rated by the Lebanese government were 57.1% compared to 42.9% working for four star hotels. As for property affiliation, the majority worked for chain hotels (61.8%) with the rest working for independents (38.2%). Finally, 78.7% of respondents classified hotels they work in as city hotels compared to 21.3 describing hotels they work in as resort hotels.

4.2 Frequency Distribution for LPI Survey

Worth mentioning that respondents in their assessment of all selected statements, allocated ten scales ranging from 1: Almost never; to 5: Occasionally; and then 10: Almost always. For each of the five practices, hotel managers responded to six items related to their perception of the frequency of use with which they used these items.

Model the way practice. *Model the way* encompasses two commitments that revolve around clarifying values by finding one's own voice and affirming shared ideas, as well as setting an example by aligning actions with shared values. In this practice, the mean for each of the six behaviors ranged from 7.75 (sets a personal example of what is expected) to 8.44 (follows through on promises and commitments).

Inspire a shared vision practice. *Inspire a shared vision* practice encompasses two commitments that revolve around envisioning the future by imagining exciting and ennobling possibilities, as well as enlisting others in a common vision by appealing to shared aspirations. The mean for each of the six behaviors in this practice ranged from 7.48 (talks about future trends influencing our work) to 8.13 (speaks with conviction about meaning of work).

Challenge the process practice. *Challenge the process* practice encompasses two commitments that revolve around searching for opportunities by seizing the initiatives, as well as experimenting and taking risks by generating small wins and learning from experience. The mean for each of the six behaviors in this practice ranged from 7.41 (experiments and takes risks) to 8.10 (makes certain that goal plans and milestones are set.

Enable others to act practice. *Enable others to act* practice encompasses two commitments that revolve around fostering collaboration by building trust, as well as strengthening others by increasing self-determination and developing competence. In this practice, the mean for each of the six behaviors, ranged from 7.39 (supports decisions other people make) to 8.67 (treats people with dignity and respect).

Encourage the heart practice. *Encourage the heart* practice encompasses two commitments that revolve around recognizing contributions by showing appreciation to individual excellence, as well as celebrating the values and victories by creating a spirit of community. The mean for each of the six behaviors in this practice ranged from 7.66 (recognizes people for commitment to shared values) to 8.48 (give team members appreciation and support).

4.3 Frequency Distribution for Managerial Competency Survey

Part B of the survey was the *Managerial Competency* Survey and contained 60 *Managerial Competencies* for hotel managers or supervisors to rate regarding importance of use. Based on Sandwith's (1993) *Competency Domain Model*, the competencies were placed in three *Domains*:*Conceptual/Creative, Interpersonal, and Leadership.* In this study an adaptation of Koenigsfeld (2007) resulted in three *Domain*s remaining the same. Each competency was rated on importance using a five point Likert scale where five (5) was critically important and one (1) was not important.

Conceptual/Creative Domain. For the *Conceptual/Creative Domain*, hotel managers responded to twelve items (eleven items plus one overall item for the *Domain*) related to their perception of the importance of use with which they used these items. The *Conceptual/Creative Domain* encompasses items related to the organization's mission, strategic environment, new idea development, and adaptation to changing circumstances (Koenigsfeld, 2007). The mean for this *Domain* ranged from 3.70 (Considers a broad range of factors "internal and/or external trends" when solving problems and making decisions) to 3.94 (Translates business strategies into clear objectives and tactics).

Interpersonal Domain. For the *Interpersonal Domain*, hotel managers responded to seventeen items (sixteen plus one overall item for the *Domain*) related to their perception of the importance of use with which they used these items. The *Interpersonal Domain* encompasses competencies related to communication functions such as writing, speaking, and listening as well as interactions with others including negotiations and providing feedback to employees (Koenigsfeld, 2007). The mean for this *Domain* ranged from 3.54 (Engages in social and causal-rapport communication with individuals outside the organization) to 4.05 (maintains working relationships and good communication with all departments).

Leadership Domain. For the *Leadership Domain*, hotel managers responded to 31 (30 plus one overall item for the *Domain*) related to their perception of the importance of use with which they used these items. In addition to providing a link between the *Conceptual/ Creative* and other*Domain*, the *Leadership Domain* encompasses competencies that link thoughts and ideas into action (Sandwith, 1993). Leaders must be trustworthy role models that exhibit trust in followers and supporters as well as empower them to do their job effectively (Koenigsfeld, 2007). The mean for this*Domain* ranged from 3.69 (Employs a team approach to solve problems when appropriate) to 4.18 (treats people with respect).

4.4 Reliability and Validity of the Measurement Scales

The survey was composed of three parts which included two scales that were used in this study. Part A was developed by the researcher and contained questions related to the *Demographic Profile* of managers where managers worked. Part A had a total of thirteen dichotomous or multiple choice items. Part B was the *Managerial Competency* survey by Koenigsfeld (2007), and adapted for this study, measured hotel managers' perceived importance of competencies. Part B had a total of 60 questions under three overall *Domain*s or *Cluster*s. Lastly, part C of the survey was the*Leadership Practices Inventory* (LPI), developed by Kouzes and Posner (2002), measured hotel managers' perceived importance of the 30 statements that make up the LPI.

Prior to answering research questions and testing hypotheses, reliability and validity analyses were conducted on each of the two scales to ensure the adequacy of their psychometric qualities. Exploratory factor analysis and internal consistency reliability analysis using Cronbach's alpha for both scales was used.

Cronbach's alpha for the three *Domains/Cluster*s of the managerial competency survey ranged from .878 (*Conceptual Creative Cluster*), followed by .894 (Interpersonal Domain) to .947 (*Leadership Domain*) which exceeded the minimum of .7 (Lee et al., 2005). For all three *Domains/Cluster*s, there were no items which would significantly increase Cronbach's alpha if deleted.

Cronbach's alpha for the five practices of *Exemplary Leadership* ranged from .821 (Model the Way) to .868 (Encourage the Heart) which exceeded the minimum of .7 (Lee et al., 2005).

4.4.1 Exploratory factor analysis for the managerial competencies survey

Factor analysis is used to identify groups or *Cluster*s of variables by reducing data from a group of interrelated variables to a smaller set of factors (Field, 2009, p. 628). By extracting interrelated variables, factor analysis ensures an instrument's construct validity (Morgan, Leech, Gloeckner, & Barrett, 2011). The *Managerial Competencies* Survey is composed of 60 statements that are grouped under three main *Domains* (Sandwith, 1993):*Conceptual/Creative, Interpersonal, and Leadership*.

Kaiser-Meyer-Olkin (KMO) and Bartlett's test of Sphericity were used to examine the sampling adequacy of items and the multivariate normality of items (Field, 2009). In general, KMO test values between 0.7 and 0.8 are considered "good", values between 0.8 and 0.9 are considered "great", and values above 0.9 are considered "superb" (Field, 2009, p. 659). As for Bartlett's test, it should have a significance value less than .05 (*p<.05*) for factor analysis to be appropriate (Field, 2009, p. 660). In this study, KMO test result was .894 and Bartlett's test had a significance value of .000. Both tests indicated that factor analysis on the scale was appropriate.

In the following step, Exploratory Factor Analysis (EFA) was conducted to examine the 60 variables that make up the *Managerial Competencies* Survey and determine which ones were associated with which *Domain* or *Cluster*. To further establish construct validity of the *Managerial Competencies* Survey, exploratory factor analysis (EFA) using Varimax rotation was conducted on the 60-item *Managerial Competencies* scale. Factor extraction was initially based on eigenvalues greater than 1.0. However, initial extraction resulted in 31 factors. Subsequent factor extraction was based on a fixed number of factors to extract, resulting in four factors that accounted for 52.30% of the total variance explained.

Generally, a loading of 0.4 is recommended for satisfactory EFA analysis (Field, 2009). This was the cutoff used when computing the factor totals for each of the Managerial Competency *Domains*/*Clusters*. The factor loadings were as follows:

Factor 1 (*Leadership/Interpersonal*) consisted of 23 items (19 *Leadership* and four *Interpersonal*) with factor loadings ranging from .406 to .639. Factor 2 (*Conceptual/Creative*) consisted of eleven items that all belonged to the same *Domain* with factor loadings ranging from .447 to .661. Factor 3 (*Interpersonal*) consisted of six factors (five *Interpersonal* and one *Leadership*) with factor loadings ranging from .422 to .583. Factor 4 (*Leadership*) consisted of four factors that all belonged to this *Domain* with factor loadings ranging from .414 to .589.

4.4.2 Exploratory factor analysis for the LPI

The LPI is composed of thirty statements where six statements measure each of the five key practices of exemplary leaders: *model the way, inspire a shared vision, challenge the process, enable others to act* and *encourage the heart*.

In this study, KMO test was .942 and Bartlett's test had a significance value of .000. Both tests indicate that factor analysis on the scale is appropriate.

In the following step, EFA was conducted to examine the thirty variables that make up the LPI and determine which ones were associated with which practice.

To further establish construct validity of the LPI scale, exploratory factor analysis (EFA) using Varimax rotation was conducted on the 30-item LPI scale. Five factors: *model the way, inspire a shared vision, challenge the process, enable others to act*, and *encourage the heart* were expected to emerge from the analysis. Factor extraction was based on eigenvalues greater than 1.0 resulting in four factors that accounted for 65.30% of the total variance explained.

Generally, a loading of 0.4 is recommended for satisfactory EFA analysis; however, to ensure all items loaded on to a factor, a cutoff of 0.35 is used (Field, 2009, p. 645). The factor loadings were as follows: Factor 1 consisted of ten items (four encourage, three enable, two model, and one challenge) with factor loading ranging from .460 to .805. Factor 2 consisted of nine items (two inspire, two enable, two challenge, two encourage and one model) with factor loadings ranging from .470 to

.749. Factor 3 consisted of five items (three inspire, one model and one challenge) with factor loadings ranging from .498 to .724. Factor 4 consisted of six items (two model, two challenge, one inspire and one enable) with factor loadings ranging from .449 to .760.

4.5 Results for Hypotheses Testing

4.5.1 Research Hypothesis 1

H1: There is a significant positive relationship between perceived importance of *Managerial Competencies* and *Exemplary Leadership* practices of managers in the Lebanese lodging industry.

To test Hypothesis 1, multiple regression analyses using the hierarchical (forward) method were performed to determine whether there was a significant explanatory (correlational) relationship between each of the *Exemplary Leadership* referred to five practices (*model the way, inspire a shared vision, challenge the process, enable others to act,* and *encourage the heart*) as essential for success as an effective manager as dependent variable and perceived importance of the three different *Managerial Competencies (1-Conceptual/Creative Domain; 2-Interpersonal Domain; 3-Leadership Domain)* as independent variables.

Before performing regression analysis, Pearson *R* was used to determine the statistical significance of the correlations between the pairs of variables. Then regression model analysis was fully performed. Worth mentioning that Collinearity statistics were examined to assure validity of the exercises performed. The Variance Inflation Factor (VIF) is a predictor of strong linear relationships with other predictors and may be a concern if over 10, while tolerance should be greater than .10 (Field, 2005, p. 224). For all the models produced in H1 regressions, the (VIF) ranged from 1.000 to 4.291, while the tolerance ranged from .240 to 1.000. These results were well within the recommended guidelines, suggesting multicollinearity was not a problem for H1 hypotheses.

Finally, although *t*-tests are easiest to conceptualize as measures of whether the predictor is making a significant contribution to the model, the standardized beta values (*β*) provide a better insight into the importance of a predictor in the model. Therefore, the standardized beta values (*β*) will be used to indicate the degree of importance in the best model (Field, 2009, p. 239).

Pearson *R* was used to determine the correlations between perceived importance of *Managerial Competencies (1-Conceptual/Creative Domain; 2-Interpersonal Domain; 3-Leadership Domain).*and the 5 practices of exemplary leadership. Results showed significant correlations with all *Managerial Competencies*. Results are presented in Table 1.

Table 1: Pearson R Correlations between the Three Managerial Competencies and Exemplary Leadership Five Practices

Managerial Competencies and Model the Way Practice	*Pearson R*	*p*-value
Conceptual/Creative	.362	.000
Interpersonal	.444	.000
Leadership	.522	.000
Managerial Competencies and Inspire a Shared Vision Practice	*Pearson R*	*p*-value
Conceptual/Creative	.357	.000
Interpersonal	.424	.000
Leadership	.483	.000

Managerial Competencies and Challenge the Process Practice	*Pearson R*	*p*-value
Conceptual/Creative	.368	.000
Interpersonal	.455	.000
Leadership	.526	.000
Managerial Competencies and Enable Others to Act Practice	*Pearson R*	*p*-value
Conceptual/Creative	.374	.000
Interpersonal	.440	.000
Leadership	.528	.000
Managerial Competencies and Encourage the Heart Practice	*Pearson R*	*p*-value
Conceptual/Creative	.362	.000
Interpersonal	.461	.000
Leadership	.543	.000

Note. $p < .05$.

Results of the regression analysis are presented next for each hypothesis.

H1a: There is a significant positive relationship between perceived importance of *Managerial Competencies* and the *model the way* practice of managers in the Lebanese lodging industry.

Using hierarchical multiple regression testing for a relationship between perceived importance of the three *Managerial Competencies* and the *model the way* practice of managers in the Lebanese lodging industry. The model had significant *F* value (F = 12.345, p = .000) with *Adjusted* R^2, explaining between 31% to 33.7% of the variance. As such, the model was selected as the explanatory model to predict managers' perceived importance of*Managerial Competencies* and the *model the way* practice.

Analysis of the individual predictors in the explanatory model indicated significant explanatory relationships between one predictor and the *model the way* practice. The standardized beta coefficient (β) for the predictor indicated its relative importance in explaining the *model the way* practice results. *Leadership* was the most important predictor in the model (t = 5.339, p = .000, β = .578). Thus, H1a was partially supported.

H1b: There is a significant positive relationship between perceived importance of *Managerial Competencies* and the *inspire a shared vision* practice of managers in the Lebanese lodging industry.

A model was produced from the hierarchical multiple regression testing for a relationship between perceived importance of the *Managerial Competencies* and the *inspire a shared vision* practice of managers in the Lebanese lodging industry. The model had significant *F* value (F = 9.441, p = .000) with *Adjusted* R^2, explaining between 25.0% and 28.0% of the variance. As such, the model was selected as the explanatory model to predict managers' perceived importance of *Managerial Competencies* and the *inspire a shared vision* practice.

Analysis of the individual predictors in the explanatory model indicated significant explanatory relationships between one predictor and the *inspire a shared vision* practice. The standardized beta coefficient (β) for the predictor indicated its relative importance in explaining *model the way* practice results. *Leadership* was the most important predictor in the model (t = 3.683, p = .000, β = .415).

Leadership Managerial Competencies had a positive relationship with the *inspire a shared vision* practice indicating that managers with higher perceptions of the importance of those *Managerial Competencies* used the *inspire a shared vision* practice more frequently. Thus, H1b was partially supported.

H1c: There is a significant positive relationship between perceived importance of *Managerial Competencies* and the *challenge the process* practice of managers in the Lebanese lodging industry.

A model was produced from the hierarchical multiple regression testing for a relationship between perceived importance of the *Managerial Competencies* and the *challenge the process* practice of managers in the Lebanese lodging industry. The model had significant *F* value, (F = 11.035,p = .000) with *Adjusted* R^2, explaining between 28.5% and 31.3% of the variance. The model was selected as the best explanatory model to predict managers' perceived importance of *Managerial Competencies* and the *challenge the process* practice.

Analysis of the individual predictors in the best explanatory model indicated significant explanatory relationships between one predictor and the*challenge the process*practice. There is no need to indicate the order of strength relationship since *Leadership* was the only significant predictor in the model (t = 4.817, p = .000, β = .535).

Leadership Managerial Competencies had a positive relationship with the *challenge the process* practice indicating that managers with higher perceptions of the importance of those *Managerial Competencies* used the *challenge the process* practice more frequently. Thus, H1c was partially supported.

H1d: There is a significant positive relationship between perceived importance of *Managerial Competencies* and the *enable others to act* practice of managers in the Lebanese lodging industry.

A model was produced from the hierarchical multiple regression testing for a relationship between perceived importance of the *Managerial Competencies* and the *enable others to act* practice of managers in the Lebanese lodging industry. The model had significant *F* value, (F = 11.264, p = .000) and had the *Adjusted* R^2, explaining between 28.9% and 31.7% of the variance. As such, this model was the best explanatory model to predict managers' perceived importance of *Managerial Competencies* and the *enable others to act* practice.

Analysis of the individual predictors in the explanatory model indicated significant explanatory relationships between two predictors and the *enable others to act* practice. The standardized beta coefficient (β) for each of the predictors indicated its relative importance in explaining the *enable others to act* practice results. *Leadership* was the most important predictor in the model (t = 5.143, p = .000, β = .565).

Leadership Managerial Competencies had a positive relationship with the *enable others to act* practice indicating that managers with higher perceptions of the importance of those *Managerial Competencies* used the *enable others to act* practice more frequently.

H1e There is a significant positive relationship between perceived importance of *Managerial Competencies* and the *encourage the heart* practice of managers in the Lebanese lodging industry.

A model was produced from the hierarchical multiple regression testing for a relationship between the ten *Managerial Competencies* and the*encourage the heart* practice of managers in the Lebanese lodging industry. The model had significant *F* value, (F = 12.599, p = .000) and had*Adjusted* R^2, explaining between 31.4% and 34.1% of the variance. As such, this model was selected as the explanatory model to predict managers' perceived importance of *Managerial Competencies* and the *encourage the heart* practice.

Analysis of the individual predictors in the explanatory model indicated significant explanatory relationships between one predictor and the *encourage the heart* practice. The standardized beta coefficient (*β*) for the predictor indicated its relative importance in explaining *encourage the heart* practice results. *Leadership* was the most important predictor in the model (t = 5.222, p = .000, β = .563).

Leadership Managerial Competencies had positive relationships with the *encourage the heart* practice indicating that managers with higher perceptions of the importance of those *Managerial Competencies* used the *encourage the heart* practice more frequently. Thus, H4e was partially supported.

5 DISCUSSION AND CONCLUSION

The importance of competency models lies in their ability to accurately develop job profiles as well as in performance appraisals by rating an employee's level of competency against a standard model or profile selected by the organization (Mirabile, 1997). Sandwith (1993) added that the value of a *Competency Domain Model* is particularly important for large organizations because it provides an excellent template that guides training programs in profiling job competencies and identifying needed competencies for each managerial level.

Large organizations with multiple managerial levels are great beneficiaries of *Managerial Competencies* (Katz, 1955); however, the hospitality industry in general and the lodging industry in particular have plenty to benefit from adopting competency models in identifying competencies necessary for current and aspiring leaders of lodging establishments. In today's global business environment, lodging companies that want to stay competitive must identify the competencies required not only for today but also for tomorrow's industry leadership. Existing research in competency modeling, while in its early stages, is already proving to be fruitful for many hospitality organizations. Prior studies in the hospitality industry, however, explored only competencies needed by lodging managers or leadership styles required for successful leadership in the lodging industry independently. This study was the first to combine theories of managerial competency modeling and a currently popular leadership style, namely, *Exemplary Leadership*. Furthermore, based on the researcher's literature review, this study was the first in Lebanon to use the *Leadership Practices Inventory* (LPI) by Kouzes and Posner. Therefore, this study extended previous research to a new area of study as well as a new culture and thus contributed to the existing body of knowledge.

The purpose of this non-experimental and correlational (explanatory) study is to determine if there is a relationship between Lebanese lodging industry managers' perceived importance of *Managerial Competencies* and *Exemplary Leadership* practices.

5.1 Research Hypothesis 1

H1: There is a significant positive relationship between perceived importance of *Managerial Competencies* and *Exemplary Leadership* practices of managers in the Lebanese lodging industry.

Hypothesis 1 tested for a relationship between the perceived importance of *Managerial Competencies (1-Conceptual/Creative Domain; 2-Interpersonal Domain; 3-Leadership Domain)* and the five practices of the dependent variable, *Exemplary Leadership (1-model the way, 2-inspire a shared vision, 3-challenge the process, 4-enable others to act,* and *5-encourage the heart*). Therefore, Hypothesis 2 had five separate sub-hypotheses.

H1 was supported. All of the five sub-hypotheses (H1a, H1b, H1c, H1d, and H1e) were partially supported indicating that there is a relationship between perceived importance of *Managerial Competencies* and each of the five practices of *Exemplary Leadership*. Although this study's literature review did not find previous research that studied the relationship between *Managerial Competencies* and *Exemplary Leadership* practices instruments, studies using similar constructs supported findings. Table 2 presents a summary of results for H1 and its sub-hypotheses. The analysis of results follows.

Table 2. H1 Summary Results

	Hypotheses	Conceptual Creative	Interpersonal	Leadership	Results
H1	There is a significant relationship between the perceived importance of*Managerial Competencies* and*Exemplary Leadership*practices of managers in the Lebanese Lodging Industry.				Partially Supported
H1a	*Managerial Competencies → Model the Way*			√	Partially Supported
H1b	*Managerial Competencies → Inspire a Shared Vision*			√	Partially Supported
H1c	*Managerial Competencies → Challenge the Process*			√	Partially Supported
H1d	*Managerial Competencies → Enable Others to act*			√	Partially Supported
H1e	*Managerial Competencies → Encourage the Heart*			√	Partially Supported

Leadership Managerial Competencies had positive relationships with the *model the way, inspire a shared vision, challenge the process, enable others to act,* and *encourage the heart* practices indicating that managers with higher perceptions of the importance of *Leadership Competencies* used all five practices more frequently. Findings were supported by study results by Tas et al. (1996) who stressed the importance of modeling the way and coaching in both training and education programs. Chathoth and Olsen (2002) study results on *Leadership* characteristics of lodging firms in India reported that challenging the process championed by their president who is designated as Chief Disorganizer is a core value for good *Leadership*characteristics. Chathoth and Olsen (2002) study results on *Leadership* reported that empowerment and enabling employees though education, training and skill development as well as rewarding and recognizing employees (similar to *encouraging the heart*) are core values for good *Leadership*characteristics. They further reported that providing a clear vision was important for incorporating change into the business strategy and the organizational structure of the company.

Furthermore, Chung-Herrera et al. (2003) study results indicated that fostering motivation (similar to encourage the heart), developing others (similar to enable others to act), and embracing change (similar to challenge the process) scored high under their *Leadership* factor. Kay and Russette (2000) study identified role modeling as essential competencies common to more than one functional area or level of management. Mwendia (2006) study results on lodging industry in Kenya and the East African Region indicated that personal role modeling, having a clear shared vision, empowerment, and enabling others to act as part of people *Leadership* competencies showed moderate positive relationship between leadership and profitability on rooms (POR). Mwendia (2006) further indicated that empowerment and enabling others to act showed moderate positive relationship between leadership and profitability on rooms (POR). Solnet, Kralji, Kay, and DeVeau (2008) study results on expectations of industry professionals from students upon internship completion reported that role modeling competencies topped the list of *Leadership Domain*.

To summarize the results for the research question of this study, the following is presented. Managers of the Lebanese lodging industry who had higher perceptions of the importance of *Leadership Managerial Competencies* used the *model the way*;the *inspire a shared vision*; the *challenge the process*; the *enable others to act*; and the *encourage the heart* practice more frequently.

The aim of this study was to examine the relationship among the *Managerial Competencies* and *Exemplary Leadership* practices of managers in the Lebanese lodging industry. Findings of this study supported the hypothesis, therefore confirming the relationship between the independent variables and dependent variable used in this study.

5.4 Implications

In the world of business management, futurists are predicting that the next 25 years will bring more change than what the world has seen in the last 100 years. If their predictions turn out to be true, *adaptation* to *innovation* must progress with the same speed of this anticipated change (Kipp, 2001). In order to achieve this goal, leaders must be prepared to face the leadership challenges that the future holds. For the lodging industry in Lebanon, hospitality leaders have a particularly daunting task in preparing current and future leaders to successfully resume or assume the managerial responsibilities that lie ahead. The implications for this study are as follows:

Hospitality educators have agreed on the need to have an educational balance that combines theoretical education with practical experience and training (Okeiyi et al., 1994). This study supports these recommendations. Both educators and industry professionals from the private sector should coordinate efforts with relevant offices from the public sector in order to provide the Lebanese hospitality industry with a labor force that is competent in managing hospitality related operations.

Similar to what Tas et al. (1996) proposed, Human Resource departments in the Lebanese lodging industry should consider adding competencies required for management positions in their hotels, in addition to job descriptions. Later on, supervisors can rate manager trainees on a checklist of competencies performed and most importantly, manager trainees can evaluate themselves against the checklist to identify areas of improvement.

This study will help Lebanese hospitality curricula in developing programs for entry level management positions that would better prepare graduates to meet current industry needs.

Leadership Managerial Competencies was perceived as the most important in relation to LPI since *Leadership Competencies* indicated a positive relationship with all five practices of *Exemplary Leadership*. Therefore, stakeholders should focus on and enrich such skills in HR practices and educational curricula.

5.5 Conclusion

In facing the rising demand for improved quality, reduced costs, and constant innovation; the need to continuously upgrade knowledge, skills, and abilities is essential for all leadership roles including those held by executives, managers, and supervisors (Dalton, 1997; Kay & Russette, 2000; Mirabile, 1997; Okeiyi et al., 1994; Perdue, Ninemeier & Woods, 2002; Sandwith, 1993; Tas, 1988; Tas et al., 1996). With management scholars racing to meet this demand, *competency modeling* has become one of the most frequently used techniques to help organizations profile jobs, select, as well as prepare the right person for the right leadership position (Katz, 1955; Sandwith, 1993). According to Sandwith (1993), the best solution can be found in creating a comprehensive competency model that delineates a distinct hierarchy of knowledge, skills, and abilities (referred to as competencies by Sandwith), required for effective managers (p.43).

5.6 Limitations

The focus on Lebanon narrowed the scope of the sample geographically to may possibly be interpreted as a convenience sampling plan. Therefore, results may not be generalizable beyond the specific population from which the sample was drawn since each country has its unique culture that may encompass economic, political, religious, and traditional values

The results of this study are meant to represent *Managerial Competencies* to *Exemplary Leadership* practices of managers in the lodging industry within the country of Lebanon. Hence, results may not be applicable to other industries, because each industry has its own leadership style, organizational culture, levels of job satisfaction, working hours, and reward system.

The study does not take into consideration any unpredictable internal or external factors, such as financial difficulties, political difficulties, or war; nor macro-economic factors or indicators such as unemployment rate, Gross Domestic Product (GDP), or Consumer Price Index (CPI).

Other profile variables, not studied in this research, and which need to be addressed are the professional and organizational profiles.

5.7 Recommendations

This study was the first to provide a research on whether a relationship between *Managerial Competencies* and *LPI* practices exists in the lodging industry in Lebanon. Several recommendations are suggested as follows:

Lebanese educators must continue to keep abreast of industry expectation and incorporate them into hospitality management curricula, as competences requirements change, curricula must be updated accordingly.

Hospitality educators should conduct a longitudinal follow up study to assess whether their graduates actually demonstrate the competencies expected by general managers and hotel companies in Lebanon.

Since the Ministry of Tourism is studying updated rules and regulations, it would be interesting to duplicate this study under the new laws. The results of determining if there is a relationship between *Managerial Competencies* and *Exemplary Leadership* practices could then be compared in both studies to observe if any differences occur, especially those related to *Administrative* competencies.

Regardless of whether changes by the Ministry of Tourism are incorporated, this study is considered a snapshot in time and therefore, must be replicated over time to determine whether competencies perceived importance have changed, or whether competencies need to be added or eliminated.

The same study could be conducted in other industries such as the banking industry or the education industry to observe if there are any relationships between *Managerial Competencies* and *Exemplary Leadership* practices. Results may reveal similarities or differences that can have national repercussions on both industry and educational preparations.

The same study could be conducted in other countries in the MENA (Middle East and North Africa) region with the same sample of the same positions in order to further investigate the relationship between *Managerial Competencies* and *Exemplary Leadership* practices. This would help in extending the body of knowledge in this field tremendously as recommended by Hodgetts (as cited in Amunckmanee, 2002) who indicated that although*Leadership* is widely recognized as a significant study of international management, however, little has been done to systematically study and compare *Leadership* approaches throughout the world since most studies have focused on specific countries or geographic areas.

Future studies could be reversed to include the input of subordinates in order to test how they observe the relationship between *Managerial Competencies* and *Exemplary Leadership* practices especially those related to *Leadership* competencies which revealed a positive relationship when it comes to *Leadership Managerial Competencies* and all five practices of *Exemplary Leadership*.

Finally, it is important to repeat the research to see if there is a relationship between the managers' demographic, professional and organizational profiles and the managerial competencies.

ACKNOWLEDGEMENT

The author would like to acknowledge the constructive criticism and editing performed by Dr. Hussin Jose Hejase, Dean of the Faculty of Business and Economics at the American University of Science and Technology.

REFERENCES

1. Baum, T. (1990), 'Competencies for hotel management: Industry expectations of education', *International Journal of Contemporary Hospitality Management,* 2(4), pp. 13-16.

2. Bornstein, S, & Smith, A. (1996), 'The puzzles of leadership'. In F. Hesselbein, M. Goldsmith, & R. Beckhard (Eds.), *The leader of the future* (pp. 281-292). San Francisco, CA: Jossey-Bass.

3. Burns, J. M. (1978), *Leadership*. New York, NY: Harper & Row.

4. Chathoth, P. & Olsen, M. (2002), 'Organizational leadership and strategy in the hospitality industry', *Journal of Services Research,* 2(1), pp. 5-29.

5. Chung-Herrera, B. G., Enz, C. A., & Lankau, M. J. (2003), 'Grooming future hospitality leaders: A competencies model', *Cornell Hotel and Restaurant Administration Quarterly,* 44(3), pp. 17-25.

6. Creswell, W. (2005), *Educational research: Planning, conducting, and evaluating quantitative and qualitative research* (2nd edition), Upper Saddle River, NJ: Pearson Education, Inc.

7. Demographics. (2010, February 9), '*Wikipedia, The Free Encyclopedia*', (Online). Available: http://en.wikipedia.org/ w/ index.php?title=Demographics&oldid=342963776. [Accessed 10 February 2014].

8. Dubrin, A. (2007), *Leadership* (5th edition), Boston, MA: Houghton Mifflin Company.

9. Field, A. (2009), *Discovering statistics using SPSS* (3rd edition), Thousand Oaks, CA: SAGE Publications Inc.

10. Goldsmith, M. (2007, September 26), 'Becoming a leader', *Business Week (Online),* p. 1.

11. Green, S. (1991), 'How many subjects does it take to do a regression analysis?', *Multivariate Behavioral Research,* 26(3), pp. 499-510.

12. Hejase, A.H. & Hejase, H.J. (2013), *Research Methods: A Practical Approach for Business Students*, (2nd edition), Philadelphia, PA: Masadir Inc.

13. Huber, D. L., Maas, M., McCloskey, J., Goode, C. J., & Watson, C. (2000), 'Evaluating nursing administration instruments', *Journal of Nursing Administration,* 30(5),pp. 251-272.

14. Katz, R. (1955), 'Skills of an effective administrator', *Harvard Business Review*, 33, pp. 33-42.

15. Kay, C., & Russette, J. (2000), 'Hospitality-management competencies', *Cornell Hotel and Restaurant Administration Quarterly,* 41(2), pp. 52-63.

16. Kerlinger, F., & Lee, H. (2000), *Foundations of behavioral research* (4th edition), Belmont, CA: Cengage Learning.

17. Koenigsfeld, J. (2007), *'Developing an industry specific managerial competency model for private club managers in the United States based on important and frequently used management competencies'* (Doctoral dissertation), Auburn University, United States--Alabama. Available: Dissertations & Theses: Full Text. (AAT 3301890).

18. Kouzes, J., & Posner, B. (1987), *The leadership challenge: How to get extraordinary things done in organizations*,San Francisco, CA: Jossey-Bass.

19. Kouzes, J., & Posner, B. (1995), *The leadership challenge: How to keep getting extraordinary things done in organizations* (2nd edition), San Francisco, CA: Jossey-Bass.

20. Kouzes, J. M., & Posner, B. Z. (2003), *Leadership practices inventory [LPI]*. (3rd edition), San Francisco, CA: Jossey-Bass, Pfeiffer.

21. Lebanon. (2010, March 16), *'Wikipedia, The Free Encyclopedia'*, [Online]. http://en.wikipedia.org/w/index.php?title=Lebanon&oldid=350281521. [Accessed 17 March 2013]

22. McClelland, D. (1973), 'Testing for competence rather than for "intelligence"', *American Psychologist*, 28(1), pp. 1-14.

23. McCrimmon, M. (2004), 'Kouzes and Posner on leadership-A critique', *Self Renewal Group,* pp. 1-6.

24. Mwendia, K. M. (2006), *'Relationship of transformational leadership competencies to financial performance: As applied to the hospitality industry'*, (Doctoral dissertation), Capella University. *ProQuest Dissertations & Theses*, 191-n/a. [Online]. Available:http://search.proquest.com/docview/30490842?accountid=((304908742). [Accessed March 12, 2014].

25. Mirabile, R. J. (1997, August), 'Everything you wanted to know about competency modeling', *Training & Development,* 51(8), pp. 73-77.

26. Nahavandi, A. (2009), *The art and science of leadership* (5th edition), Upper Saddle River, NJ: Pearson Prentice Hall.

27. National College for School Leadership [NCSL]. (2003), *'The leadership challenge James Kouzes and Barry Posner'*. [Online]. Available:http://www.ncsl.org.uk/media/F7B/97/randd-leaders-business-kouzes.pdf. [Accessed 15 May 2015]

28. Okeiyi, E., Finley, D., & Postel, R. T. (1994), 'Food and beverage management competencies: Educator, industry, and student perspective', *Hospitality & Tourism Educator*,6(4), pp. 37-40.

29. Peg, T. (1996), 'The leadership challenge: How to keep getting extraordinary things done in organizations', *Human Resource Development Quarterly,*7(3), pp. 300-303.

30. Posner, B, Z., & Kouzes, J.M. (1993), 'Psychometric properties of the leadership practices inventory', *Educational and Psychological Measurement,*(53),pp. 192-198.

31. Powers, T., & Barrows, C.W. (2006), *Introduction to management in the hospitality industry* (8th edition). Hoboken, NJ: John Wiley & Sons, Inc.

32. Republic of Lebanon (2010), *'Syndicate of Hotel Honors in Lebanon'*. [Online]. Available: Retrieved on February 9, 2010 fromhttp://

www.informs.gov.lb/cms.informs.gov.lb/Cultures/en-US/Menu/Directory/Private+Sector/. [Accessed 9 February 2014]

33. Ryan, B., Tavitiyaman, P, & Weekarit, N. (2009), *'The impact of gender, age and education related to leadership competencies needed for success as a hotel general manager'*, Paper presented at the 2009 International CHRIE Conference-Refereed Track, University of Massachusetts, Amherst, MA. [Online]. Available: http://scholarworks.umass.edu/cgi/viewcontent.cgi?=1175&context=refereed. [Accessed March 9, 2014]

34. Rutherford, D. (1987), 'The evolution of the hotel engineer's job', *Cornell Hotel and Restaurant Administration Quarterly,* 27(4), pp. 72-78.

35. Sandwith, P. (1993), 'A hierarchy of management training requirements: The competency domain model', *Public Personnel Management,* 22(1), pp. 43-62.

36. Sapienza, D. (1978), 'What university hotel students ought to study: Opinions expressed by a selected group of Nevada hotel executives', *Journal of Hospitality Education,* 2(2), pp. 11-16.

37. Sherwood, S., & Williams, G. (2009, January 11), The 44 Places to Go in 2009. *New York Times,* (final edition), p. 3. [Online]. Available: NewsBank on-line database (Access World News). [Accessed 23 March 2013]

38. Solnet, D., Kralji, A., Kay, C., & DeVeau, L. (2008), 'A lodging internship competency model: Enhancing educational outcomes through work integrated learning', *Journal of Hospitality & Tourism Education,* 21(4), pp. 16-24.

39. Sumner, M., Bock, D., & Giamartino, G. (2006), 'Exploring the Linkage between the characteristics of IT project leaders and project success',*Information Systems Management*23(4), pp. 43-49.

40. Tajfel, H., & Turner, J. C. (1979), 'An integrative theory of intergroup conflict', In S. Worchel & W. G. Austin (Eds.), *The social psychology of intergroup relations* (pp. 33-47). Monterey, CA: Brooks/Cole.

41. Tas, R. F. (1988), 'Teaching future managers', *Cornell Hotel and Restaurant Administration Quarterly,* 29(2), pp. 41-43.

42. Tas, R. F., LaBrecque, S. V., & Clayton, H. R. (1996), 'Property-management competencies for management trainees', *Cornell Hotel and Restaurant Administration Quarterly,* 37(4), pp. 90-96.

43. The Leadership Challenge. (n.d.), *'The LPI'*. [Online]. Available: http://eu.leadershipchallenge.com/WileyCDA/Section/id-131089.html. [Accessed 20 June 2014]

44. Media.Wiley. (2002), *'The leadership practices inventory: Theory and evidence behind the five practices of exemplary leaders: Appendix0512BP'*. Available: The Leadership Challenge website [Online]. http://media.wiley.com/assets/463/74/lc_jb_appendix.pdf . [Accessed 23 March 2014]

45. Tourangeau, A., & McGilton, K. (2004), 'Measuring leadership practices of nurses using the Leadership Practices Inventory', *Nursing Research,* 53(3), pp. 182-189.

46. Work, J. (1996), 'Leading a diverse work force', In F. Hesselbein, M. Goldsmith & R. Beckhard (Eds.), *The leader of the future* (pp. 281-292), New York: Jossey-Bass Inc.

47. World Travel & Tourism Council [WTTC]. (2010), *'Travel & Tourism Economic Impact Lebanon'*. [Online]. Available: http:// www.wttc.org/bin/pdf/original_pdf_file/lebanon.pdf. [Accessed 23 March 2014]

Chapter 4

Decision support for risks managers in the case of deliberate food contamination: The dairy industry as an example

Beate Pinior, Franz J. Conraths, Brigitte Petersen, Thomas Selhorst

HIGHLIGHTS

- Evaluation of the consequences of a deliberate contamination of the milk supply chain by mathematical modeling;
- Application of optimization algorithms for food defense issues in the context of Enterprise Risk Management (ERM);
- Development of a contingency plan for the most probable entry sources, which can be used as possible starting points for control measurements in the operational-systems of the dairies;

ABSTRACT

Dairy farms were identified, which can be included in a contingency plan set up to prevent or mitigate the consequences of deliberate contamination of a food supply chain. The deliberate introduction of a contamination into the supply chain of milk was simulated in a scenario where milk producers serve as the entry sources and consumers of milk represent the target to be affected by the contamination. It is shown that the entry sources have an impact on the damage caused, i.e. in terms of the number of consumers reached. A contingency plan is provided that contains a list of entry sources ranked according to their impact on the damage to consumers. To generate this list, a computer program was developed that simulates the impact of the contaminations on consumers via the trade of contaminated milk. Possible variations in the trade links between milk producers, dairies and consumers as well as between dairies are considered. It is investigated how these trade links alter the generated list of entry sources.

The results indicate that, regardless of the actual milk trade flow, control measures should be introduced on 39% of the milk producers in order to minimize the damage. The identification of suitable entry sources may help risk managers to focus on these farms in a contingency plan that improves the sensitivity of control activities related to deliberate contamination.

KEYWORDS

Greedy algorithm; Terrorist attacks; Contingency plan; Dairy industry

1. INTRODUCTION

Risks in supply chains can come from a large number of sources [1], [2], [3], [4], [5], [6],[7], [8] and [9] and thus their prevention leads to a broader view of risk management [10]. For example in the study of Wu and Olson [11], financed risks of enterprises were addressed and models were applied in order to support their investment decision-making[11]. Other studies estimated the risks of a food contamination in order to provide the risk-informed decision-making on food safety management issues [12] and [13]. However, Enterprise Risk and Enterprise Risk Management (ERM) have attracted a great deal of attention, especially in recent years [14]. But there are slightly different views on ERM[14] and [15]. In general, ERM is defined as a systematic and integrating approach to managing all risk factors which an organization is faced with [15] and thus represents the most effective way for companies to manage or mitigate their risks [10]. Besides [15], COSO [16] defines ERM as "a process [that is] effected by an entity's board of directors, management and other personnel, applied in strategy setting and across the enterprise, designed to identify potential events that may affect the entity, and manage risk to be within its risk appetite, to provide reasonable assurance regarding the achievement of entity objectives". Similarly, Olson and Wu [17] define ERM as "the integrated process of identification, analysis and either acceptance or mitigation of uncertainty in investment decision making". Against this theoretical background, our research framework rests on the definition of risk management provided by [17] with a focus on the decision-making of supplier selection under consideration of their vulnerability to terrorist entry sources. The food supply chains are tempting targets for terrorists as attacks on these systems may destabilize the economy and disrupt the flow of foods [18]. Defense preparedness in this field is often in the hand of the private sector [19]. For example, companies in the food sector apply for certifications of their food defense management strategies [20]. Rasco and Bledsoe [19] claimed that about 80% of consumers consider the food supply as vulnerable to attacks. Several incidents of intentional contamination in the food supply chain underline its vulnerability [21], [22], [23], [24] and [25]. For instance, at least 751 people were affected due to deliberate contamination of salad bars in Oregon, USA, in 1984 by members of a religious commune [25], [26] and [27]. Another case occurred in 2003, where approximately 100 people were affected after consumption of ground beef that had been contaminated by a supermarket employee [25] and [28]. According to Sobel et al. [29], an intentional contamination of the food supply may be similar to an accidental contamination. In this context, the likely size of damage caused by an attack can be inferred from observed unintentional foodborne disease outbreaks [29]. In 1994, for example, a large outbreak of *Salmonella enteritidis* in the United States affected approximately 224,000 people after accidental contamination of pasteurized liquid ice cream [22], [29] and [30].

However, the above mentioned studies are based on the assumption that the selection of the entry sources for a deliberate contamination of the supply chain can be random, because the consequences of deliberate of accidental contaminations cannot be distinguished from each other.

In this paper, we concentrate on the hypothetical threat posed by a deliberate introduction of a pathogen or toxins into the milk supply chain. We focus on a scenario, where the milk producer (dairy farm) is used as the entry source for a contamination and where milk consumers are the target of the attack [13]. We assume that a potential attacker would aim at reaching a maximum spread of the contaminated milk and at using a minimum number of milk producers as entry sources for the contamination. Ideally, the attacker would aim at reaching the maximal spread of contaminated milk by contaminating the first milk producer in the network of milk trade. If the attacker was not stopped after the first assault, he would contaminate another milk producer as a second entry source if this contamination cause larger increase of infected consumers compared to the first entry sources. The attacker has achieved his goal, when all consumers have been supplied with the contaminated milk. Due to the fact that the milk trade between milk producers, dairies and consumers as well as between dairies in the milk supply chain is dynamic [31] and [32], we hypothesize that trade links may influence the selection of milk producers that are used as the entry sources for the contamination by the attacker.

However, the most important task during a foodborne outbreak is to identify the source of the food contamination and its entry sources [33]. Thus, the aim of this paper is to provide the following information for risk managers on the chosen scenario: Firstly, which entry sources would be chosen by a hypothetical attacker, if data on the commodity flows became publicly available? Secondly, how many entry sources would the attacker have to choose to reach all consumers with contaminated milk in Germany? Thirdly, in which sequence would an attacker choose potential entry sources? Fourthly, are there milk producers who can be selected independent of the flow of milk to induce maximum damage? Fifthly, in the context of ERM, what strategies can be derived to prevent or mitigate the consequences of deliberate contamination with scant resources? These questions were answered by proposing a contingency plan.

To prevent that the results of this research are used as an instruction for a potential attacker, we work with highly aggregated and anonymous data. Moreover, we use a random gravity model to generate the trade connections between the actors in the milk supply chain. Furthermore, we focus on the spread of hypothetically contaminated milk via trade. Our investigation does take any features into account that are specific for particular milk producers, such as bio-security measures. Nevertheless, we expect that contamination of milk in dairy plants is less likely than in farms due to restricted access to the dairy plants. Relevant characteristics of milk, the kind of biological agents or toxins, individual dispositions like the age of people [34], [35], [36] and [37] and internal processes like pasteurization [38], [39], [40], [41], [42] and [43], which may influence the vulnerability of the consumers to contamination, but also the spread of contamination[32], are not considered in this paper.

2. MATERIAL AND METHODS

2.1. The generation of the milk trade network

The underlying milk trade network has been described in detail elsewhere [32]. In brief, the term "milk trade network" comprises the trade connections between milk producers (*P*) and dairies (*D*), dairies and consumers (*C*) and the trade connections between dairies. On the one hand, the horizontal flow of milk between dairies (inter-dairy trade) and on the other hand, the vertical flow (without inter-dairy trade) between milk producers, dairies and consumers is taken into account. All milk producers of a country and all consumers of a municipality were aggregated into one milk producer node or one consumer node, respectively. The milk trade network consists of 12,597 nodes, with$P$=294, D=80 and C=12,223.

Data on the trade relations from milk producer nodes to dairy nodes are available in accordance with the German law of market regulations for goods (Marktordnungsmeldeverordnung). Information on the trade connections between dairy nodes and consumer nodes as well as between dairy nodes was not available and these trade connections were predicted through a standard randomized gravity model [32]. The standard randomized gravity model [44], [45], [46] and [47] was based on the assumption that the probability of two market actors trading with each other is proportional to the supply and demand of the respective actors and indirectly proportional to their distance to each other [32], [46] and [48]. Further information on the generation of the German milk trade network can be found in [32]. However, different trade networks are required due to the random nature of the model [32]. One hundred different trade networks were therefore created, consisting of 50 trade networks with inter-dairy trade and 50 without inter-dairy trade.

2.2. Greedy algorithm and objective functions

To identify the number and the rank-order of milk producer nodes, which may cause maximum damage in terms of the number of contaminated consumer nodes, the greedy algorithm was used. This algorithm can solve optimization problems [49], [50] and [51]and is applied under the predetermined objective function to find the most appropriate milk producer nodes (*P*) as entry sources for a contamination to cause maximum damage on the condition that maximum spread of contaminated milk in association with a maximum number of contaminated consumer nodes (*C*), so that the number of milk producer nodes involved in spreading the contamination is minimal (Eq. (1)).

equation(1)

$$\max |\{c : c \in C;\ \min|\ \{p : p \in P\}\ |, p \in D, c \in D, D \in R\}|$$

In this context, there is a second condition requiring that trade connections between milk producer nodes and dairy nodes ($p \in D$) as well as between dairy nodes and consumer nodes ($c \in D$) exist. Furthermore, the condition should be reflexive and transitive (*R*), as trade connections between dairy nodes should be considered in our model. However, the objective function considers only the trade volume (*v*) and the trade connections of the milk producer nodes, their associated dairy nodes and the consumer nodes.

294 candidate of entry sources were hypothetically contaminated in the computer simulations, selected and sorted according to the extent of the resulting damage caused dependent on the respective milk trade flow.

The greedy algorithm starts with the identification of the candidate set of solutions. A candidate is selected for the solution when it maximizes the selection function. Let *S*represent the ordered set of selected candidates and *si* is a member of this ordered set at position *i*. Each member of the set is assigned a value $\Delta v(si)$. This value represents the additional weighted number of contaminated consumer nodes, when the milk producer node *s* is added to the set of solutions. The weight ($wi=i^{-1}$) depends on the position (*i*) of the milk producer node in the ranked list of solutions ($i=1,\ldots, n$). With respect to the objective function, a milk producer node is only added to the list if $S_{i+1}>S_i$ with $S_i = \sum_i^n \Delta v(s_i\ w_i)$. Therefore, the greedy algorithm searches for the "best" milk producer node according to the number of contaminated consumer nodes, sets this milk producer node on the first position in the list of potential entry sources and then follows with the "next-best" milk producer node (Fig. 1). As a consequence, every place in the list can be filled by only one milk producer node. The greedy algorithm cannot handle the theoretical case where two milk producer nodes have exactly the same value and should therefore occupy the same place in the list of potential entry sources.

The algorithm stops when *Si* cannot be improved any more or, in other words, when all consumer nodes (*C*) are contaminated (Fig. 1). The scenario, in which all consumers are reached, represents a worst-case situation. The greedy algorithm was applied for two kinds of milk trade networks, one with inter-dairy trade and one without. We therefore obtained two contingency plans over 50 simulations for each kind of the milk trade network. One simulation represents one sub-trade network.

Step (1): Contamination of a milk producer node and calculating the number of consumer nodes that can be hypothetically contaminated (absolute); Step (2): Selecting the milk producer nodes that have reached the maximum number of consumer nodes. The milk producer node with the maximum number of infected consumer nodes was used as a baseline for further evaluations of milk producer nodes; Step (3): Select the second-best milk producer node, on the condition that it has infected other consumer nodes than the first milk producer node (maximum increase for the first milk producer node); Step (4): Sorting the milk producer nodes according to extent of damage leading to a rank-order of the milk producer nodes in the contingency plan.

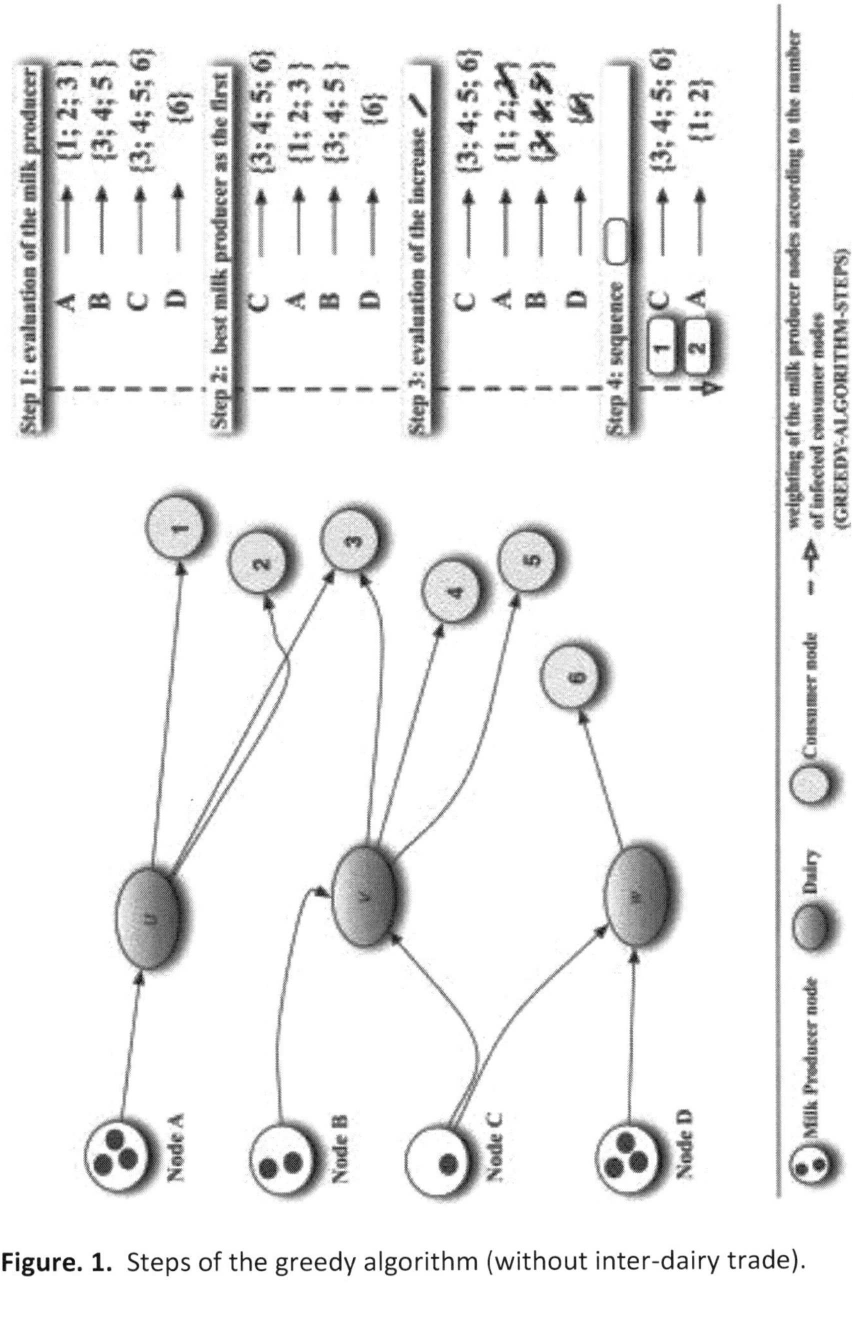

Figure. 1. Steps of the greedy algorithm (without inter-dairy trade).

To answer the question, whether milk producer nodes can be used regardless of the flow of milk as appropriate entry sources for a contamination, a comparison between the two contingency plans according to the most commonly observed portals of entry was conducted. To achieve a better comparison between the milk producer nodes, which may occur in both contingency plans, and those, which are only included in one of the contingency plans, the impact of these milk producer nodes for the damage situation was determined with Eq. (2).

equation(2)

$$I_{(p)} = \sum_{i=1}^{r} h(r)g(r)$$

The impact of milk producer nodes $I(p)$ is calculated using its rank (r) (position) in the ordered contingency plan determined by the greedy algorithm for each of the considered random networks. Let $h(r)$ be the frequency of (p) at position (r) using all simulation results, where $g(r)$ is a weight, $g(r)=(r+1)-i$. The value of r is determined by the greedy algorithm and resembles the maximum size (length of the contingency plan) of the best-ordered milk producer nodes in random network considered.

3. RESULTS

The contingency plan includes the number and rank-order of milk producer nodes for inducing the maximum damage situation as well as the associated number of contaminated consumer nodes (Fig. 2 and Fig. 3) for the two kinds of trade networks (with and without inter-dairy trade). In Fig. 2 and Fig. 3(b), each line represents a simulation process of a total of 50 simulations. Each line represents the number of milk producer nodes which are involved in the worst-case situation per simulation. Furthermore, each column shows the rank of the milk producer nodes for the worst-case situation according to the reached number of consumer nodes. Fig. 2 and Fig. 3(a) depict the damage situation in respect to hypothetically contaminated consumer nodes in association with the rank-order of the milk producer nodes.

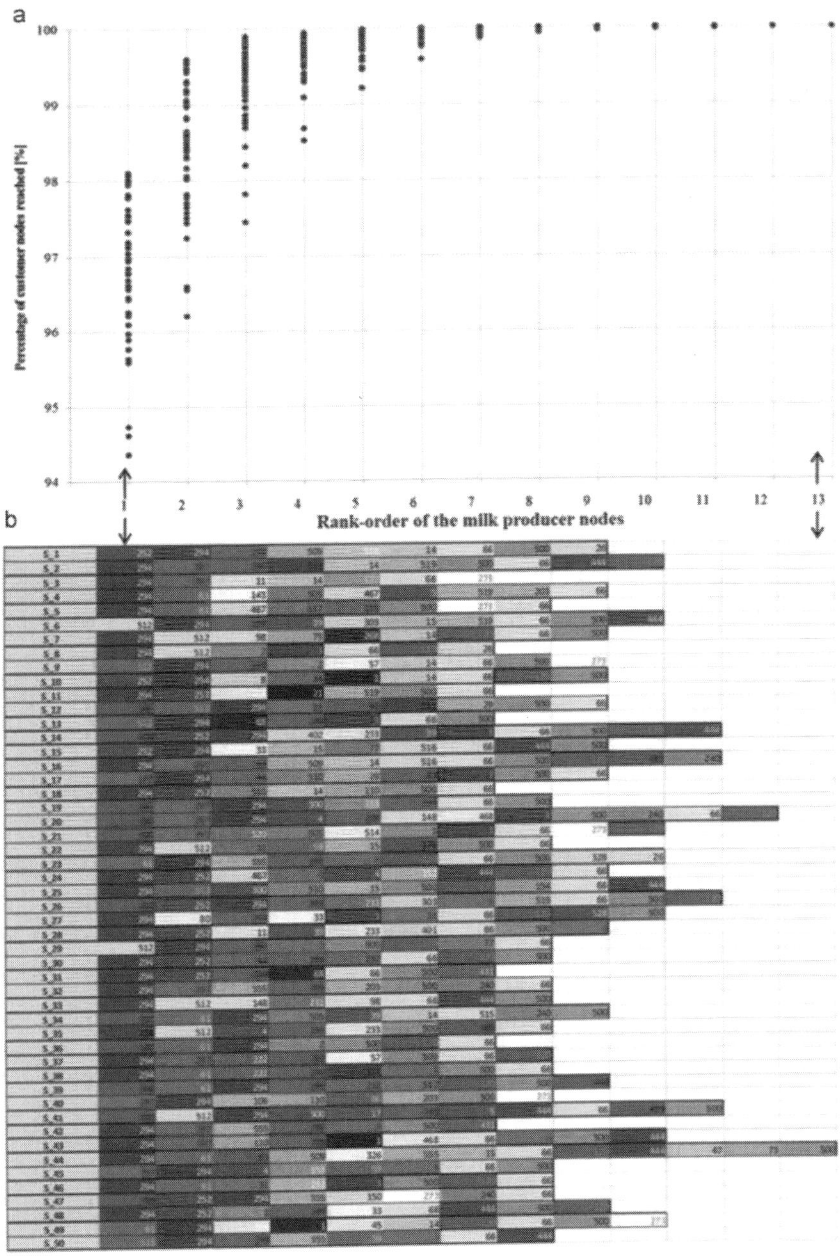

Figure. 2. Contingency plan taking inter-dairy trade into account in association with the hypothetically contaminated consumer nodes. (a) Size of the damage situation with respect to the number of hypothetically contaminated consumer nodes per rank-order of the milk producer nodes. Each point represents the damage caused by a milk producer node in the underlying column of 2b. (b) Results of a total of 50 simulations. Each line shows the number of milk producer nodes that are involved in the worst-case situation per simulation; each column shows the rank-order of the milk producer nodes for the worst-case situation according to reached number of consumer nodes. The milk producers on the first rank caused the maximum damage situation. The milk producer nodes on the second place caused the largest increase of infected consumer nodes compared to the first milk producer nodes.

When the inter-dairy trade structures are taken into account, the entire contingency plan includes 86 (29.2%) different milk producer nodes from a total of 294 (100%). For all 50 simulations, the greedy algorithm calculated that the minimum required number of milk producer nodes to achieve the worst-case situation was seven and that the maximum number of milk producer nodes was 13 (Fig. 2(b)). On average, by the introduction of a contamination into nine milk producer nodes a worst-case situation can be introduced. More than 94% of the consumer nodes could be contaminated upon introduction of the hypothetical pathogen or toxins into the milk producer nodes on the first rank of the contingency plan (Fig. 2(b)). The milk producer nodes on the second rank led to a further increase of the damage situation of 1–2% (Fig. 2(a)). All subsequent milk producer nodes induce a minimal increase in the number of contaminated consumer nodes by 1– 4%.

One milk producer node (ID: 294) was identified in 49 out of 50 simulations (Fig. 2(b)). Two milk producer nodes (ID: 66 and 500) showed up in 89% of all simulations. The remaining 83 milk producer nodes appeared on average 3.6 times in the simulations. In all simulations, 27 different milk producer nodes appeared only once (Fig. 2(b)).

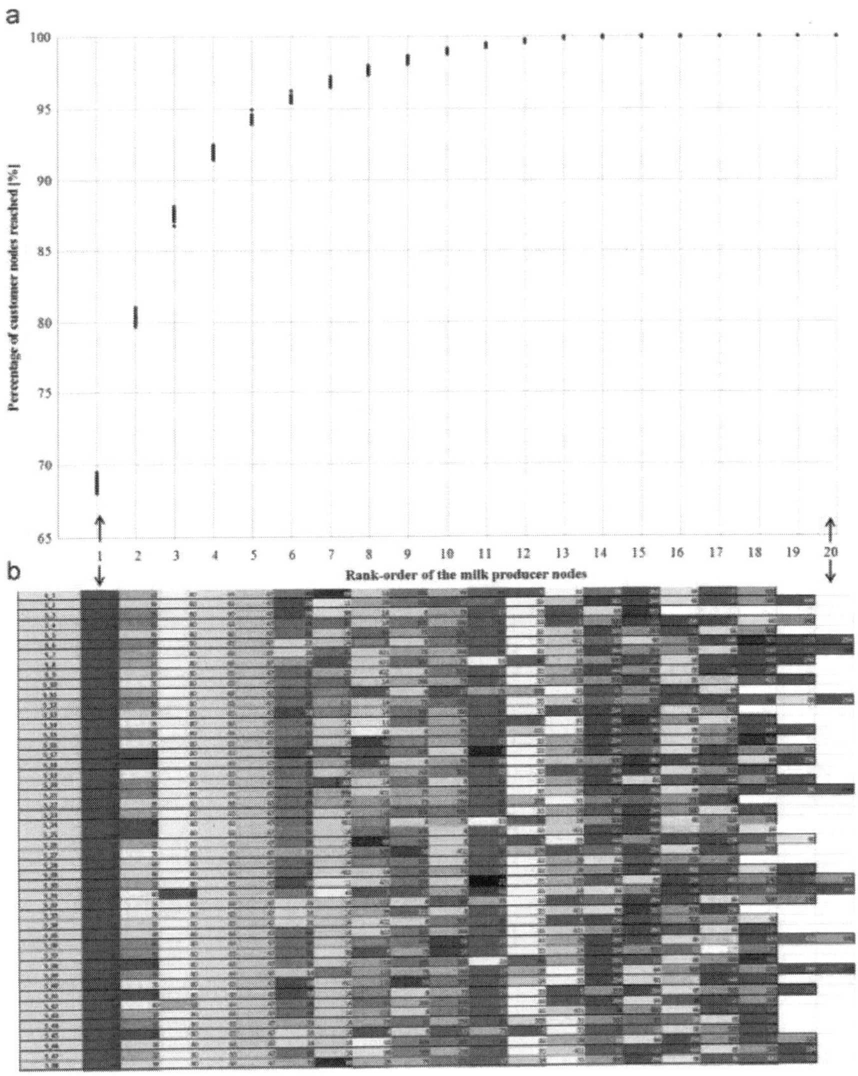

Figure. 3. The contingency plan without inter-dairy trade in association with the hypothetically contaminated consumer nodes. (a) Size of the damage situation with respect to the number of hypothetically

contaminated consumer nodes per rank-order of milk producer nodes. Each point represents the damage caused by a milk producer node in the underlying column of 3b. (b) Results of 50 simulations. Each line shows the number of milk producer nodes that are involved in the worst-case situation per simulation; each column shows the rank-order of the milk producer nodes for the worst-case situation according to reached number of consumer nodes. The milk producers on the first rank caused the maximum damage situation. The milk producer nodes on the second place caused the largest increase of infected consumer nodes compared to the first milk producer nodes.

One milk producer node (ID: 294) was on the first rank in 26 of 50 simulations (Fig. 2(b)). In all remaining simulations, this milk producer node ranked second (*n*=13) or third rank (*n*=10) in the contingency plan. Such "direct neighborhood" ranking of milk producer nodes in the contingency plan was observed for 68 (79.1%) nodes in all simulations (Fig. 2(b)).

If the inter-dairy trade is neglected, a worst-case situation could be caused on average by 18.4 (minimum 15, maximum 20) contaminated milk producer nodes (Fig. 3(b)). The contingency plan contained a total of 76 (25.8%) different milk producer nodes (Fig. 3(b)). The milk producer nodes with the highest rank in the contingency plan could infect more than 68% of all consumer nodes (Fig. 3). The milk producer nodes on the second rank led to a maximal further increase of the damage situation of 12.2% (Fig. 3(a)). In contrast to the results for scenario with the inter-dairy trade, we always found the same milk producer node (ID: 100) on the first rank in all simulations in the scenario without inter-dairy trade (Fig. 3(b)). Moreover, three milk producer nodes were found in all 50 performed simulations, but on the ranks 2–5. All other 72 milk producer nodes were identified ten times on average in the simulations on the ranks 2–20.

A "direct neighborhood" ranking of milk producer nodes in the contingency plan without inter-dairy trade was identified for 63 (82.9%) nodes in all simulations (Fig. 3(b)).

The contingency plan without inter-dairy trade contained 29 milk producer nodes that were not included in the plan that took inter-dairy trade into account (Fig. 4(b)). These milk producer nodes were on average 7.7 times included in all simulations and were between ranks 2 and 19 in the rank-order. Conversely, 39 milk producer nodes existed in the contingency plan with inter-dairy trade that were not included in the plan without inter-dairy trade (Fig. 4(b)). These milk producer nodes had ranks between 1 and 9 in the rank-order and appeared on average 3.4 times in the simulations. 64% of these milk producer nodes were involved twice at the maximum in all performed simulations.

In order to determine the impact of the milk producer nodes for a contingency plan, we used the maximum number of participating milk producer nodes (r=20) as obtained from the greedy-algorithm calculation (Fig. 3(b)). In this context, the maximum impact ($I_{max}(p)$) of a milk producer in the contingency plan is 2000, if the milk producer node is identified on the first rank (r_{max}=20) in all 100 simulations. Fig. 4(a) shows the common intersection of the both contingency plans. 47 (29.0%) of the milk producer nodes were present in both plans and therefore relevant regardless of the milk flow. The maximum impact of a milk producer node in the contingency plan was 1220 and the minimum 14. The average impact of the milk producer nodes was 260. If the impact of the common milk producer nodes in the contingency plan is compared separately for the trade flows with and without inter-dairy trade, it becomes clear that the milk producer nodes that have a high impact in the trade network without inter-dairy trade do not have the same level of importance for the trade flows with inter-dairy trade (Fig. 4(a)). No clear correlation (R=0.013) existed between these milk producer nodes regarding their impact on a damage situation. In this context, the milk producer nodes in the plan without inter-dairy trade had a 33.5% higher impact on the damage than the nodes in the list with inter-dairy trade.

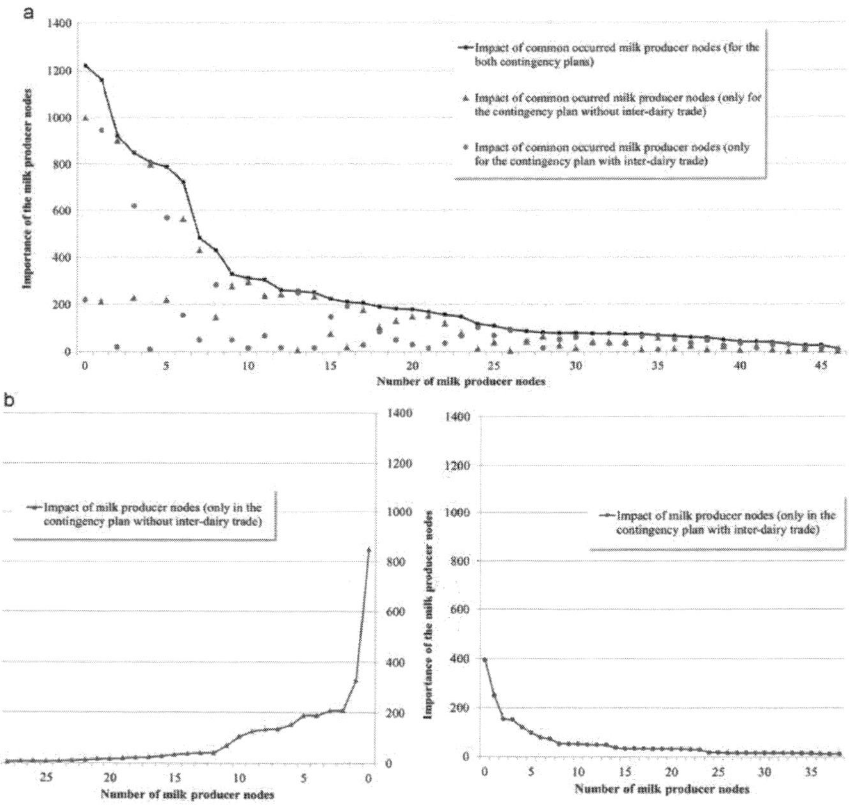

Figure. 4. Impact of milk producer nodes, (a) present in both contingency plans or (b) in only one of the contingency plans.

Fig. 4(b) shows that the milk producer nodes in the contingency plan with inter-dairy trade possess on average half of the impact for a damage situation as the milk producer nodes in the contingency plan without inter-dairy trade.

4. DISCUSSION

The assessment of the consequence of a deliberate or accidental release of a contamination in a food supply chain can only be done by computer simulation. The simulation results obtained in this study are only valid for the scenario described in the introduction and summarized in Eq. (1).

Confidential handling of data on supply structures related to deliberate contamination is important. This implies to realize the balancing act between creating knowledge for risk managers and to avoid creating instructions for an attack. On the one hand, the information provided for risk managers should answer the question "what should be done?" [52] in case of an attack or a threat of an attack. On the other hand, risk managers who use the results of simulation should understand the strengths and weaknesses of simulation results to avoid incorrect conclusions [53].

The strengths of simulations result from the fact that new insights for risk managers are provided, which cannot be offered by other means.

Our results showed that 60% of the milk producer nodes are not suitable as possible entry sources in our model because only 25.8% (without inter-dairy trade *n*=76) and 29.2% (with inter-dairy trade *n*=86) of the milk producer nodes as entry sources were identified in a total of 294 possible entry sources for each contingency plan. 29% of the milk producer nodes were present in both contingency plans and therefore relevant regardless of the milk flow. Especially the milk producer node with the ID 100 proved as important for risk managers because this milk producer node was found in 66 out of 100 simulations as the "best" milk producer node for a maximum spread of hypothetically contaminated milk. This milk producer represents a suitable entry source regardless of the flow of milk. Our findings indicate that cost- and time-efficient identification of such milk producer nodes (dairy farms) is possible if detailed trade data are used in the simulation. This may help risk managers to identify critical points with regard to the entry sources for a contamination in a food supply chain.

In addition, our research has proven that different trade structures have a significant impact on the number and the rank-order of selected entry sources for contaminations. Our hypothesis, that the number and the rank-order of milk producer nodes as entry sources for a contamination may vary due to the different trade flows of milk (Fig. 2 and Fig. 3(b)), was confirmed and can be underpinned by the following facts: First, the number of milk producer nodes as portals of entry for a pathogen to induce a worst-case situation varies within and between the contingency plans per simulation (min: 7|13; max: 15|20). The entry sources that exist regardless of the kind of trade, i.e. with or without inter-dairy trade (Fig. 4(a)), have a very different impact on the damage situation. It is not possible to tell that milk producer nodes identified as important in one of the contingency plans are also important in the other plan or vice versa. The simulations showed that trade between the dairies leads to a significant change of the potential entry sources for a contaminations. If the trade of milk between dairies is taken into account (Fig. 2(b)), more diversity at potential entry sources per rank-order becomes visible as compared to the list without inter-dairy trade (Fig. 3(b)). This observation can be underpinned by the fact that there were 50% more potential entry sources for contaminations on the first 13 ranks compared to the list of portals of entry for the network without inter-dairy trade (Fig. 2 and Fig. 3(b)). The consequence of this large number of entry sources for contaminations per rank is that no general statement can be made, which allows to select milk producer nodes that need to be monitored to mitigate the maximum damage. One consequence of a non-specific selection of the control points would be a delay in coping with the damage situation. In this context, a potential attacker runs a higher risk to select a milk producer node that is not relevant for the spread of the contamination as the selection is highly dependent on the respective milk trade flow, in contrast to the contingency plan without inter-dairy trade (Fig. 3(b)). Three milk producer nodes (ID: 100; 63; 47) appeared always on the same rank and on the first ranks in all simulations without inter-dairy trade. This means that a potential attacker could reach almost all consumer nodes via these entry sources. Milk producer nodes on the first rank can reach 68% of the consumer nodes (Fig. 3). The importance of these milk producer nodes as entry sources can be explained by the fact that these milk producer nodes delivered their milk to different dairy nodes. In a previous study it was shown that milk producer nodes delivered their milk on average to three different dairy nodes whereas some milk producer nodes delivered to up to 20 different dairy nodes [31]. Consequently, more trade connections in the milk trade network existed. If an attacker would choose a milk producer node from the first rank in the contingency plan with inter-dairy trade, 94% of consumer nodes could be reached. The importance of the milk producer nodes for the spread of the contamination in the inter-dairy trade network was shown by [32]. It was calculated that the spread of contaminated milk through some milk producer nodes could be higher if trade between dairy nodes existed. Generally, the importance of the inter-dairy trade for the spread of a contamination is illustrated by the fact that 30% of the total milk production in Germany is traded between dairies [31]. A detailed description of the German milk supply chain can be found in [31].

One of the limitations of the model results from the fact that the number of milk producer nodes not involved in the spread of a contamination in our model can be less than 60%, because the greedy algorithm cannot consider the theoretical case that two milk producer nodes may cause exactly the same extent of damage. In contrast to the model of Wein and Liu [13], our model does not include microbiology, processing, time-resolved delivery structures and compliance with existing security measures such as International Food Standard (IFS), British Retail Consortium (BRC) or Hazard Analysis and Critical Control Points (HACCP) etc., which may already exist at different levels of the milk supply chain and could influence the dimension of potential damage as well as the selection of the potential entry sources for contaminations. Another limitation is that the hypothetical attacker must process the complete data set to plan the attack beforehand and implement the scenario, which is unlikely in reality. A further limitation of this study is that the gravity model used to generate the trade network was only based on three variables (supply, demand and distances between actors), although it is well known that additional variables, such as the price of traded goods [44] are also important for the decision whether or not trade relations between actors take place [32]. An overview of the economic factors, which are essential for the choice of supplier partners in dairy industry, is given by [54]. Apart from economic factors, effects of terrorism on supplier selection require more attention [55]. Whereby, selection of supplier is a multi-criteria decision making [55], [56], [57], [58] and [59].

However, it has been shown by this study that for the decision makers the type and number of trading partners can have a significant impact on the range of a damage situation. Furthermore, the suppliers, which should find more consideration in the operational-control-system of the dairies, were illustrated by means of the contingency plan. This information can influence the decision making in the supplier selection process. The surveillance activities for selected scenarios can be reduced by half by decision makers, if current data on trade flows are present. Nonetheless, it was shown in this work that an increased transparency or a high availability of data about the supply structure can lead to a maximal damage with minimal efforts based on the selection of suitable entry sources for a potential attacker. The protection of supplier data is essential in order to avoid such a scenario and therefore for food defense issues in the context of ERM.

Further research will focus on the kind of agencies or specific stages of the production process, which can influence the number of consumers reached, the kind of selected entry sources for a deliberate contamination and the assessment of the consequences related to deliberate contamination. This will be done with the help of the findings gained here.

ACKNOWLEDGMENT

We thank Dr. R. Carmanns, Bavarian State Ministry for Nutrition, Agriculture and Forestry (StMELF-HIT, Munich), and the Federal Office for Agriculture and Food (BLE, Bonn) for providing data. We thank all reviewers and the editor for their comments, which substantially helped to improve the manuscript. This work was supported by the German Federal Ministry of Education and Research (BMBF, Bonn), Research Grant 13N11208, as well as by the Project VET-Austria, a co-operation between the Austrian Federal Ministry of Health, the Austrian Agency for Health and Food Safety and the University of Veterinary Medicine, Vienna.

REFERENCES

1. D.L. Olson, D. Wu **Risk management models for supply chain: a scenario analysis of outsourcing to China** Supply Chain Management: An International Journal, 16 (2011), pp. 401–408
2. Aydin G., Babich V., Bei D.R.l, Yang Z., Decentralized supply risk management. Working paper. Georgetown University; 2009.
3. P.D. Berger, A. Gerstenfeld, A.Z. Zeng **How many suppliers are best? A decision-analysis approach** Omega, 32 (2004), pp. 9–15
4. C. Blome, T. Schoenherr **Supply chain risk management in financial crises—a multiple case-study approach** International Journal of Production Economics, 134 (2011), pp. 43–57
5. Oke, M. Gopalakrishnan **Managing disruption in supply chains: a case study of a retail supply chain** International Journal of Production Economics, 118 (2009), pp. 168–174
6. Sarkar, P.K.J. Mohapatra **Determining the optimal size of supply base with the consideration of risks of supply disruptions** International Journal of Production Economics, 199 (2009), pp. 122–135

7. O. Tang, N. Musa **Identifying risk issues and research advancements in supply chain risk management** International Journal of Production Economics, 133 (2011), pp. 25–34

8. C.S. Tang **Robust strategies for mitigating supply chain disruptions** International Journal of Logistics. Research and Applications: a Leading Journal of Supply Chain Management, 9 (1) (2006), pp. 33–45

9. H. Yu, A.Z. Zeng, L. Zhao **Single or dual sourcing: decision-making in the presence of supply chain disruption risks** Omega, 37 (2009), pp. 788–800

10. D. Wu, D.L. Olson, J.R. Birge **Introduction to special issue on Enterprise risk management in operations** International Journal of Production Economics, 134 (2011), pp. 1–2

11. D. Wu, D.L. Olson **Enterprise risk management: a DEA VaR approach in vendor selection** International Journal of Production Economics, 48 (2010), pp. 4919–4932

12. E. Doménech, I. Escriche, S. Martorell **Quantification of risks to consumers' health and to company's incomes due to failures in food safety** Food Control, 18 (2007), pp. 1419–1427

13. M.L. Wein, Y. Liu **Analyzing a bioterror attack on the food supply: the case of botulinum toxin in milk** Proceedings of the National Academy of Sciences USA, 102 (28) (2005), pp. 9984–9989

14. D. Wu, D.L. Olson **Computational simulation and risk analysis: an introduction of state of the art research** Mathematical and Computer Modelling, 58 (2013), pp. 1581–1587

15. G. Dickinson **Enterprise risk management: its origins and conceptual foundation** The Geneva Papers on Risk and Insurance, 26 (3) (2001), pp. 360–366

16. Committee of Sponsoring Organizations of the Treadway Commission (COSO). Enterprise risk management-integrated framework. New York; 2004. Available on: 〈http:// www.

coso.org/documents/coso_erm_executivesummary.pdf⟩ [accessed on 01.09.2014]

17. D.L. Olson, D. Wu **Enterprise risk management models**, 978-3-642-11473-1Springer, Heidelberg (2010)

18. C.G. Turvey, B. Onyango, W. Hallman, S.C. Condry **Consumers' perception of food-system vulnerability to an Agroterrorist attack** Journal of Food Distribution Research, 38 (3) (2007), pp. 70–87

19. Rasco, G.E. Bledsoe **Food defense in an aquaculture setting** Journal of the World Aquaculture Society, 41 (2) (2010), pp. 175–191

20. IFS. International Featured Standards; 2014. Available on: http://www.ifs-certification.com/index.php/de/certification-bodies-de/ifs-standards/ifs-food⟩ [accessed on 09.07.2014].

21. L. Manning, N.R. Baines, A.S. Chadd **Deliberate contamination of the food supply chain** British Food Journal, 107 (4) (2005), pp. 225–245

22. A.S. Khan, D.L. Swerdlow, D.D. Juranek **Precautions against biological and chemical terrorism directed at food and water supplies** Public Health Reports, 116 (1) (2001), pp. 3–14

23. S.A. Kolavic, A. Kimura, S.L. Simons, L. Slutsker, S Barth, C.E. Haley **An outbreak of Shigella dysenteriae type 2 among laboratory workers due to intentional food contamination** The Journal of the American Medical Association, 278 (5) (1997), pp. 396–398

24. L.A. Sobering **Food Defense preparedness in small and very small meat and poultry establishments** ([Master's thesis]) Kansas State University, Manhattan, KS (2008)

25. E. Yoon, C.W. Shanklin **Food terrorism: perceptional gaps between importance and performance of preventive measures** Journal of Foodservice Business Research, 10 (4) (2007), pp. 3–23

26. T.J. Török, R.V. Tauxe, R.P. Wise, J.R. Livengood, R. Sokolow, S. Mauvais, *et al.* **A large community outbreak of salmonellosis caused by intentional contamination of restaurant salad bars** The

Journal of the American Medical Association, 278 (5) (1997), pp. 389–395

27. Veiga **Food defense and security: the new reality** H. Alpas, S.M. Berkowicz, I. Ermakova (Eds.), Environmental security and ecoterrorism, Springer, Dordrecht, The Netherlands (2011), pp. 39–54

28. CDC (Center for Disease Control and Prevention) **Nicotine poisoning after ingestion of contaminated ground beef-michigan** Mortality Morbidity Weekly Report, 52 (18) (2003), pp. 413–416

29. J. Sobel, A.S. Khan, D.L. Swerdlow **Threat of a biological terrorist attack on the US food supply: the CDC perspective** Lancet, 359 (9309) (2002), pp. 874–880

30. W.T. Hennessy, W.C. Hedberg, L. Slutsker, E.K. White, M.J. Besser-Wiek, E.M. Moen, *et al.* **A national outbreak of *Salmonella enteritidis* infections from ice cream** New England Journal of Medicine, 334 (20) (1996), pp. 1281–1286

31. Pinior, U Platz, U. Ahrens, B. Petersen, F. Conraths, T. Selhorst **The German Milky Way: trade structure of the milk industry and possible consequences of a food crisis** Journal on Chain and Network Science, 12 (1) (2012), pp. 25–39

32. B. Pinior, M. Konschake, U. Platz, H.D. Thiele, B. Petersen, F.C. Conraths, *et al.* **The trade network in the dairy industry and its implication for the spread of contamination** Journal of Dairy Science, 95 (11) (2012), pp. 6351–6361

33. M. Ercsey-Ravasz, Z. Toroczkai, Z. Lakner, J. Baranyi

34. Complexity of the international agro-food trade network and its impact on food safety

35. PLoS One, 7 (5) (2012), p. e37810 http://dx.doi.org/10.1371/journal.pone.0037810

36. S.K. Crerar **Ensuring food safety throughout the food supply Chain** Asian Journal Animal Science, 13 (13 Suppl.) (2000), pp. 376–385

37. J.M. Liu, A Ren, L. Yang, J. Gao, L. Pei, R. Ye, *et al.* **Urinary tract abnormalities in Chinese rural children who consumed melamine-contaminated dairy products: a population-based screening and follow-up study** Canadian Medical Association Journal, 182 (2010), pp. 439–443

38. J.B. McCabe-Sellers, E.S. Beattie **Food Safety: emerging trends in foodborne illness surveillance and prevention** Journal of the American Dietic Association, 104 (2004), pp. 1708–1717

39. Rocourt J., Moy G., Vierk K., Schlundt J. The present state of foodborne disease in OECD countries; 2003. Available at: http://www.who.int/foodsafety/publications/foodborne_disease/oecd_fbd.pdf〉 [accessed on 21.11.2012].

40. K. Bake, U. Beeck, B. Dyck, A. Eyers, B. Feuerriegel, P. Gieske, *et al.* **Handbuch der Milch- und Molkereitechnik** Gelsenkirchen, Germany (2003)

41. V. Franzen **Untersuchungen zur Mikrobiologischen Qualität von Frischkäse verschiedener Herstellungsweisen** ([PhD thesis]) Justus–Liebig-University, Gießen, Germany, DE (2009)

42. Riemelt, B. Bartel, M. Malczan Milchwirtschaftliche mikrobiologie, Behr, Hamburg, Germany, DE (2003)

43. VO 853/2004: Verordnung (EG) Nr. 853/2004 des Europäischen Parlaments und des Rates vom 29. mit spezifischen Hygienevorschriften für Lebensmittel tierischen Ursprungs; April 2004.

44. O.G. Weingart, T. Schreiber, C. Mascher, D Pauly, B.M. Dorner, H.F.T. Berger, *et al.* **The case of botulinum toxin in milk: experimental data** Applied and Environmental Microbiology, 76 (2010), pp. 3293–3300

45. P. Zangerl **Mikrobiologie der Produkte** V. Krömker (Ed.), Kurzes Lehrbuch Milchkunde und Milchhygiene (2007), pp. 156–179

46. J.E. Anderson, E. van Wincoop **Gravity with Gravitas: a solution to the border puzzle** American Economic, 93 (1) (2003), pp. 170–192

47. J.H. Bergstrand **The gravity equation in international trade: some microeconomic foundations and empirical evidence** Journal: Review of Economics and Statistics, 67 (3) (1985), pp. 474–481

48. J.J. Lewer, H. van den Berg **A gravity model of immigration** Economics Letters, 99 (1) (2008), pp. 164–167

49. J.E. Anderson **The gravity model** Annual Review of Economics, 3 (2011), pp. 1–46

50. J.A. Bikker **An extended gravity model with substitution applied to international trade** P.A.G. van Bergeijk, S. Brakman (Eds.), The gravity model in international trade: advances and applications, Cambridge University Press, Cambridge (2010), pp. 135–164

51. Bouchet **Greedy algorithm and symmetric matroids** Mathematical Programming, 38 (1987), pp. 147–159

52. Hüftle, M. Heuristik. Available at: 〈http://134.169.42.157/Methoden/HeuriOpt/HeuriOpt.pdf〉 [accessed on 29.11.2012]

53. B. Korte, D. Hausmann **An analysis of the greedy heuristic for independence systems** Annals of Discrete Mathematics, 2 (1987), pp. 65–74

54. B.S. Harsh, J.L. Connor, D.G. Schwab **Managing the farm business** Prentice-Hall, Eglewood Cliffs, New Jersey (1981), p. 384

55. J.M. Keeling **Models of foot-and-mouth diseases** Proceedings of the Royal Society of London B, 272 (2005), pp. 1195–1202

56. B. Pinior, V. Belaya, B. Petersen, T. Selhorst, Structures and relationships in supply chains and networks: conceptual issues and application in German dairy sector. In: Proceedings of the conference paper at international conference on economics and management of networks. p. 1–21.

57. F.T.S. Chan, N. Kumar **Global supplier development considering risk factors using fuzzy extended AHP-based approach** Omega, 35 (2007), pp. 417–431

58. E.A. Demirtas, Ö. Üstün **An integrated multiobjective decision making process for supplier selection and order allocation** Omega, 36 (2008), pp. 76–90

59. W. Ho, X. Xu, P.K. Dey **Multi-criteria decision making approaches for supplier evaluation and selection: a literature review**

60. European Journal of Operational Research, 202 (2010), pp. 16–24 O. Ustun, E.A. Demirtas **An integrated multi-objective decision-making process for multi-period lot-sizing with supplier selection** Omega, 36 (2008), pp. 509–521

61. W. Xia, Z. Wu **Supplier selection with multiple criteria in volume discount environments** Omega, 35 (2007), pp. 494–504

Chapter 5

A CULTURAL PERSPECTIVE OF THE GENERAL MANAGERS' WORK: THE GREEK 4 & 5* HOTELS' CASE

Giousmpasoglou Charalampos

ABSTRACT

This paper explores the General Managers' (GMs) Roles and Competencies in Greek 4 & 5* hotels from a contextual perspective, focused in culture. The existing literature indicates that the Greek context influences managerial work to a certain degree. The country's participation in 2 international surveys during the 1990s (CRANET, GLOBE project) have indicated characteristics that differentiate Greek management, comparing to the so called 'western' management. In order to identify the influence of culture in the work of the GMs in 4 and 5* a hotel, a research was conducted in 16 luxury (4 & 5*) city and resort hotels in four popular destinations: Athens, Thessaloniki, Crete and Rhodes. In total 32 GMs and their assistants participated in this country case study. The results have indicated that Greek managers fully understand and appreciate the generic managerial competencies framework and roles required in any other European country. On the other hand they try to cope with the contextual challenges appearing mainly due to the Greek culture by adapting these roles and competencies to their working environment. In addition, international and national hotel chains are following standard operating procedures and formal policies which in most of the cases are adjusted or fitted in the Greek context. Smaller independent hotels (usually family business) are more informal in relation to policies and standards.

KEYWORDS

Hospitality Industry, Roles, Competencies, Culture, Greece

1. INTRODUCTION

Managerial work in Greek hotels has been a neglected and under researched area, despite the significant contribution of the industry in the Tourism sector and the Greek economy. Greek hotel General Managers (GMs) in luxury establishments (4 & 5*) shape their competencies and perform their roles required for their position, under the influence of a strong national culture. This research paper explores the influence of the Greek context in managerial work in 4 & 5* hotels.

2. LITERATURE REVIEW

2.1. The Greek Context

Greece has a long tradition in tourism and hospitality mainly due to its history and ancient civilisation. The 'modern' hospitality industry has emerged since the 1950s to cater for the tourism demand. Due to the rapid growth of the tour operators in Europe and the phenomenon of the mass tourism, Greece experienced a dramatic increase of tourism flows to in the late 1970s and 1980s, facilitated by plenty of natural, cultural and environmental resources, existing airport infrastructure in major islands, and lower cost of living in comparison with most of Europe (Eurostat, 2005). In addition Buhalis (2001) argues that Greek resorts have different product and market profiles making them capable of satisfying a great diversity of tourism demand. The Greek tourism product is an amalgam of natural, cultural and heritage attractions spread throughout the country, as well as a wide variety of services offered predominantly by small and medium tourism enterprises (SMTEs). Moreover, 15,000 miles of coastline; 2,500 islands; an average of 300 sunny days annually; a unique fauna and flora; as well as climatic superiority with mild winters and warm summers are some of its natural attractions.

Today the Hospitality & Tourism sector in Greece contributes approximately 15.00% of the National GDP ranking third in the E.U. after Spain (18.38%) and Portugal (15.40%) according to the WTTC (2005). The sector occupies totally 808,000 employees 18% of the country's entire labour force; a total of 96,750 employees work in hotels and 24,895 in small lodges (SETE, 2003).

The Geek and International literature suggests that Greek management has hardly existed until the early 1980s; all management practices and methods were largely adoption of MNCs practices. Kanelpoulos (1990) has documented a lack of wide diffusion of modern management methods and systems such as formal structures, planning and control systems, human resource management systems, incentive systems, and management information systems. Bourantas and Papadakis (1996) argue that the salient characteristics of Greek management (in the 1980s and early 1990s) were:

1 Concentration of power and control in the hands of top management.

2 Lack of modern systems to support strategic decisions.

A question that was raised here is whether Greek management possessed any unique characteristics that distinguish it from other European management styles (e.g., the institutionalised participation of employees in Germany or Sweden and the informal network relationships among small and medium-sized enterprises in Italy). The answer came during the 1990s and the early 2000s trough the participation of the country in two international surveys: the Price Waterhouse Cranfield Project (CRANET) concerning Human Resources strategies and policies across Europe (Papalexandris and Chalikias, 2002); and the GLOBE (Global Leadership and Organisational Behaviour Effectiveness) project which provided useful insights for each participative country cultural perspectives in relation to management and leadership (Javidan & House, 2001). The findings of these significant surveys indicated the country's differentiation in management practices, due to the existing socio-cultural context.

The results from the CRANET survey revealed that in Greece, as in other European countries, there is evidence of both convergence and divergence in HRM policies and the overall work context (Myloni et al., 2004). On the one hand the introduction of common legislation and agreements between countries of the E.U., will eventually lead to harmonisation of the Industrial Relations (I.R.) and H.R.M. systems across different national contexts (Brewster, 1994). In addition, globalisation forces multinational companies (MNCs) to adopt common H.R. practices in their overseas subsidiaries. On the other hand, there is still a persistent belief that social and cultural differences between countries will continue to supersede the forces of globalisation emanating from technologically driven markets or supranational agreements (Sparrow & Hiltrop, 1997). Moreover, Myloni et al. (ibid.) have found that for the range of H.R. issues examined in CRANET survey (Business Policy and Evaluation, Industrial Relation, Recruitment and Selection, Management Training and Development, Employee Benefits and Reward, Performance Evaluation) MNC subsidiaries have adapted parent company H.R.M. practices to the local ones, up to a point. These practices might be characterised by high levels of cultural susceptibility and to certain degree sensitivity to cultural differences. On the contrary Greek firms seem to adapt less H.R. 'Best Practices' and continue to follow the local patterns and norms. According to the above, it is possible that Greek firms still have some way to go in terms of facing direct competition in the global market. Myloni et al. (ibid.) conclude that Greek companies are still embedded to their cultural environment to a considerable extent.

The globe project has provided a better insight of the relation between management practices and national culture in greece. perhaps the most important finding is the country's score to the "in-group collectivism" dimension (also referred as "family collectivism") which reflects the extent to which a society's institutions favour autonomy versus collectivism. this dimension refers to the extent to which members of a society take pride in membership in small groups such as their family and circle of close friends, and the organisations in which they are employed. in greece being a member of a family and of a close group of friends, an in-group, is very important to people. papalexandris et al. (2002) indicate that one of the greek culture's main characteristics is strong family bonds, even though in big cities there might have been a recent change in this respect. the father is the centre of the family; he is responsible for all its members and the one who makes the final decision. there is a strict hierarchy and younger members are expected to show respect to the older. power is concentrated in a few hands, which is usually accepted although it does not go unquestioned. family members and close friends tend to have strong expectations from each other. taking care of their needs and satisfying their expectations is critical to each individual. it is not unusual to forego due diligence, or equal employment opportunity, and to favour a close friend or family member in recruiting or in allocating rewards and promotions. making regular references to one's family and especially one's father is quite acceptable and can go a long way in opening doors.

Despite the paternalistic family oriented management style there are indications for a strong will to change. Figures from the GLOBE project (House et al., 2002) show that Greece has low mean scores in 'society as is' and higher scores in 'society should be'; these results confirm the existence of a culture gap found also in previous research studies in Greek organisational culture. According to Bourantas and Papadakis (1996), there is a discrepancy between general organisational culture as perceived by managers and their personally preferred culture. This is considered to be an indication of the desire for change within organisations. The greatest pressures for convergence are coming from the obligations of Greece as a member of the E.U. and several other organisations which require planning ahead and efficient management of the various projects. While this affects mostly the public sector, globalisation put pressures for uniform management practices and policies in private sector organisations. Thus, a slow but steady movement towards harmonisation of management practices at least with the rest of the E.U. members is observed.

2.2. The Nature of Managerial Work

In order to understand the nature of managerial work in hotels there is a need to answer the questions 'what managers do' and 'why they do what they do'. Among the numerous efforts to establish a credible account of the managerial work (i.e. Carlson, 1951; Martin, 1956; Burns, 1957; Sayles, 1964; Stewart, 1967, 1976; Steward et al., 1980; Kotter, 1982; Luthans et al., 1985; Carroll & Gillen, 1987; Whitley, 1989) the literature reveals that only Mintzberg (1973, 1994) managed to conclude in a comprehensive and robust model. According to Mintzberg's view (ibid) the manager is working in a chaotic environment spending most of his time talking to others – in and out from the workplace – influencing any kind of people, collecting information from various sources such as gossips and rumours and generally trying to maintain a very delicate balance like jugglers do. This empirical study found managerial roles to be highly variable, involving the often simultaneous pursuit of a variety of objectives in changing ways according to the judgment of the individual manager in the particular situation.

Mintzberg grouped these ten roles into three broad categories: inter-personal (Figurehead, Leader, Liaison), informational (Monitor, Disseminator, Spokesman) and decisional roles (Entrepreneur, Disturbance handler, Resource allocator, Negotiator). His work triggered many similar studies which fell in the following paradox: although they rejected Mintzberg's Roles Model for the shake of a new model creation, one way or another came to similar conclusions with Mintzberg. Despite the original work's weaknesses (Fondas and Steward, 1994), it remained for almost three decades the only straightforward model for the nature of managerial work. Twenty years later Mintzberg (1994) provided a revised version of his work (figure 1) justifying carefully his model by covering most of the points that have been fiercely attacked in the past (Hales, 1999).

A plethora of writers attempted to develop managerial work frameworks for hospitality – in most of the cases unsuccessfully. The main reason for this failure is the involvement of many disciplines in hospitality research and the lack of valid and reliable sources. Hospitality research has been preoccupied with Mintzberg's ideas and several researchers replicated or tested his early work. Studies of managerial hospitality work have addressed three questions that have divided the work chronologically (Dann, 1990). Early research (pre-1973) was concentrated with the questions 'what managers do and how' focused very often in how they allocate their time (Nailon, 1968). The middle-period representing the time between the early 1970s and the late 1980s is pro-occupied with what managers do in terms of roles (Ley, 1980; Pickworth, 1982; Ferguson & Berger, 1984; Nebel & Ghei, 1993; Mount & Bartlett, 1999). These studies have replicated and developed the framework presented by the general studies of managerial work drawing especially from Mintzberg's early work (1973).

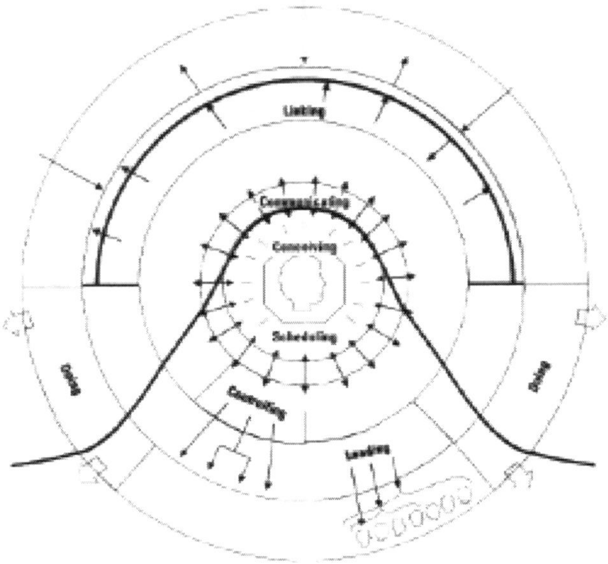

Source: Mintzberg (1994), p.23

Figure 1: Managerial Work Rounded Out

Table 1: A summary of Managerial Work Studies in the Hospitality Industry

Author/s	**Year**	**Focus of Study**
Nailon	1968	Hotel Managers Time & Contact Patterns
Ley	1980	Hotel GMs Roles and Performance
Arnaldo	1981	Hotel Managers Roles, Leadership
Ferguson & Berger	1984	Hotel Manager Roles Framework

Hales & Nightingale	1986	Framework for Hospitality Unit Mgrs' Work
Umbreit and Eder	1987	Hotel Mgrs Behaviour & Effectiveness
Wosford	1989	Hotel Managers Leadership Role
Shortt	1989	Managerial Work Survey (MWS)
Guerrier & Lockwood	1990	Literature Review in Hospitality Mgrs'
Dann	1991	Roles Literature Review in Hospitality Mgrs' Roles
Mullins & Davis	1991	Managerial Effectiveness in Hospitality
Nebel & Ghei	1993	Conceptual Framework of Hotel Mgr's Job
Kim	1994	Hotel Managers' Roles in Korea
Peacock	1995	Hospitality Manager Effectiveness
Gore	1995	Hotel Managers' Decision Making
Teare	1995	Literature Review in Int. Hospitality Mgmt.
D'Annunzio-Green	1997	Int. Hospitality Management Development
Ladkin	1999	Literature Review in Hotel Mgr Roles
Mount & Bartlett	1999	Hotel Management Roles Survey (MRAS)

Brotherton	1999	Hospitality Management Framework
Stalcup & Pearson	2001	Hotel Management Turnover Causes

The period from the late 1980s until the late 1990's has focused in managerial behaviour and performance (Eder & Umbreit, 1989; Worsfold, 1989; Mullins & Davis, 1991; Peacock, 1995; Gore, 1995). A forth period is added to Dann's (ibid) chronological categorization covers the time from the mid-1990s until today and focuses on the skills and competencies required in order to perform effectively the managerial roles (Christou & Eaton, 2000; Kay & Rousette, 2000; Chung-Herrera et al., 2003; Kay & Moncarz, 2004).

2.3. Managerial Competencies

Throughout the second half of the 20th century understanding of the 'manager' has been a matter for debate and calls for action, though managerial performance has been notoriously difficult to evaluate (Child, 1969; Anthony, 1986). The research on managerial roles could not provide credible and sufficient answers to the measurement of management performance. The competence approach that appeared in the 1980s marked a new development; its focus lies in endorsing and promoting types of managerial behaviour rather than measuring managerial outcomes.

As early as in 1973, McClelland claimed that job performance should be predicted from competence rather than intelligence; his work has triggered the 'behavioural' approach in management competencies. The term 'competence' was first used in a managerial context in the research of the McBer Consultancy in the late 1970s in the USA as part of the initiative by the American Management Association to identify the characteristics which distinguish superior from average managerial performance. The work was encapsulated in The Competent Manager (Boyatzis, 1982).

Boyatzis defined the term 'competency' as 'an underlying characteristic of a person in that it may be a motive, trait, aspect of one's self-image or social role, or a body of knowledge which he or she uses' (1982, p.21). This approach was labelled as the 'personal characteristics' or 'behavioural' approach emphasises the distinction between threshold competencies which all job holders require – the competencies necessary for someone to fill the job – and differentiating competencies which distinguish the outstanding from the average manager (Boyatzis, 1982; Spencer & Spencer, 1993).

There has been an enormous diversity of interpretation of the meaning of the term 'competence' or 'competency', and no agreed definition (Rees, 2003). Woodruffe (1993) pointed out, defining the word according to Boyatzis' definition, leaves the term open to a multitude of interpretations. To avoid unresolved debates about 'motives', 'traits' and so on, the term 'competence' can be used to refer to a 'set of behaviours, skills, knowledge and understanding which are crucial to the effective performance of a position' (Woodruffe, ibid., p. 29).

The term and its related concepts have been adapted in number of ways. It has been extended to cover the training of a select group of managers and to the total change of an entire organisation. Despite Boyatzis' original intention to provide a model of competence that could be validated against organisational criteria, competencies have also been taken up at a national level and provide the framework for example, for developing general management competences in the U.K. (Townley, 1999). In this case the Management Charter Initiative (MCI, 1991) has adopted a functional approach to competence, which reflects a greater focus on task, seeking to identify concretely the work functions which a competent manager should be capable of performing

(Cheng et al., 2003). For the purpose of this paper however, the discussion will focus in the frameworks that are falling in the 'behavioural' approach.

Table 2: Methods/Approaches for measuring performance and competence

Researcher	**Performance measure**	**Competence measure**
Boyatzis, (1982)	Supervisory nominations and ratings, work output measures	Behavioural Event Interview (BEI)
Schroder, (1989)	Work group / managerial performance	Behavioural Observation Method (BOM)
Spencer & Spencer, (1993)	Various managerial levels performance	Job Competence Assessment (JCA)
Dulewicz & Herbert, (1991-99)	Career advancement	360 degree Ratings (from supervisor, peers and subordinates)
Cheetham & Chivers, (1996-98)	Professional Performance in different professions	Provisional Model Critical Peer Evaluation / Reflection

According to Iversen (2000) all the different models within the 'behavioural' approach are primarily based on the study of the competency (competent behaviour) of outstanding performers. The major contributors within this approach are based in research conducted in the U.S. (Boyatzis, 1982; Schroeder, 1989; Spencer & Spencer, 1993), but there are also significant contributions from the U.K. (Cockerill, 1989; Dulewicz & Herbert, 1996; Cheetham & Chivers, 1996/8).

The various approaches have been encapsulated in the shape of a competency model/framework. This is a descriptive tool that identifies the knowledge, skills, abilities and behaviour needed to perform effectively in an organisation (Lucia & Lepsinger, 1999). Competency frameworks are considered to be beneficial in that they assist jobholders to contribute significantly to their personal development by enabling them to understand clearly what is required to perform effectively in a particular role, as well as in a wider context (i.e. throughout the industry). They also provide a framework within which to develop tools and techniques designed to improve performance (Brophy & Kiely, 2002). Competence frameworks and methods vary considerably from organisation to organisation and the extent and depth to which they become part of human resource functions can also differ (Rees & Garnsey, 2003).

Table 3: A comparison of managerial competency frameworks in Hospitality

Competency Areas('clusters' according to**Dulewicz & Herbert, 1999**)	**Tas (1988); Baum (1991); Christou & Eaton (2000)**	**Lockwood (1993)**	**Kay & Russette (2000)**	**Brophy & Kiely (2002)**	**Chung – Herrera (2003)**
1. Intellectual	Operational Awareness	Managing Operations & Business	Conceptual – Creative;	Planning & Organising, Problem	Industry Knowledge; Critical

			Technical	Solving	Thinking
2. Personal	Ethics; Professionalism; Legal Responsibility	Personal Management Skills		Enthusiasm	Self Management
3. Communication	Communication		Administr a-tive	Effective Communication	Communica -tion
4. Inter-Personal	Customer Problems Handling		Inter-personal	Teamwork	Inter-personal
5.Leadership	Employee Relations; Leadership; Motivation	Managing People	Leadership		Leadership
6. Results – Orientation	Development & Control of Productivity; Customer relations			Leading for results, Customer Service Focus; Financial Awareness; Strategic Thinking	Implement a-tion; Strategic Positioning
No. of Competencies:					
40	36	78	18	36	99

By examining the established competency frameworks of the behavioural approach (table 2), it can be argued that competencies typically gather in 5 'clusters': Intellectual/ information handling, Achievement /results orientation, Managing and leadership, Motivational / Interpersonal, Personal (Dulewicz & Herbert, 1999). A sixth cluster was added by the work of Cheetham & Chivers (1996, 1998) that of Values and Ethics. Although this approach has been criticised for being too general by not recognising that the competency mix may vary from position to position, it remains the most popular approach in both sides of the Atlantic.

Since the concept of managerial competencies became fashionable in the late 1980s, a number of studies have been conducted to identify the essential competencies of managers in the hospitality industry. The majority of the competency frameworks / models (table 3) developed for the hospitality industry, are falling in the behavioural approach which is concerned with superior performance. Most of these studies examine the competencies required of the hospitality graduates in the UK and US respectively. Linking effectively the Hospitality managerial work with the Hospitality Higher Education has been always an issue; regardless the approach in Hospitality Education, there was always a demand to create a skilled core of Hospitality graduates ready to cope with the diverse environment of the sector. Thus, since the early 1990s a growing number of tourism and hospitality university courses that aim to meet the demands of a volatile and changing world (Umbreit, 1993) have taken up the challenge to prepare students by developing and enhancing the management competencies and skills needed to operate successfully. This movement has been supported by the industry's growing demand of suitable qualified managerial staff, which until today is regarded as scarce.

In the late 1980s Tas (1988) carried out a study that targeted the management competences required by graduate trainees in the hotel business. This study was part of an effort to change the nature of the hospitality management curriculum which traditionally had a vocational

/ technical orientation (Baum, 2002). It involved the examination of the views of the general managers of 75 properties with 400 or more rooms. Baum (1991) has replicated the study in the UK and was based on the response of 118 hotel GMs out of 223 hotels with 150 or more rooms. The third replication of the study came surprisingly from Greece where Christou & Eaton (2000) who surveyed 178 hotels (4 & 5*) with 91 reponses from the GMs. The common finding for all three studies was that general managers identified the 'soft skills' as essential. There where however some gaps between the perceptions of Greek GMs compared to those of from the UK and the US: Greeks where very reluctant to consider any area as unimportant and rated most competencies as 'essential'. The main limitation of these studies is the methodological quantitative approach which as Eaton & Christou (ibid) suggest could be combined with qualitative tools such as in-depth interviews, in order to triangulate the data. Since most of the hospitality managerial competencies studies suffer from 'cultural and conceptual myopia', the differences that have been surfaced between the study of Tas (1988) in the US, Baum (1991) in the UK, and Christou & Eaton (2000) in Greece, proves that contextual and situational factors mater.

Under this fact lies the hart of a debate which is broadly represented by a pluralist and a unitarist approach. On the one hand Bartlett & Ghoshal (1997) argue that "situational factors vary so much that it is impossible to make a generic list of managerial competencies that are relevant for most managerial positions". On the other hand Spencer & Spencer (1993) suggest that "superior managers of all types and levels share a general profile of competencies. Managers of all types are more like each other than they are like the individual contributors they manage". It is difficult to decide which position is closer to the hospitality industry profile. This dilemma has confronted organisational studies for decades. As Ruth (2006) argues the problem of developing a competency framework involves abstraction, aggregation and standardisation. The extent to which this takes place is precisely the extent to which its applicability and usefulness in any particular situation is vitiated.

3. RESEARCH PROFILE, DISCUSSION OF THE FINDINGS AND CONCEPTUAL FRAMEWORK

3.1. The Research Profile

The research that conducted as part of a PhD Thesis, served mainly two aims: first to explore the GMs' roles and competencies profile in Greek 4 and 5* hotels; and second to investigate the role of culture and the level of influence in GMs' work.

In total 16 hotel case studies with 32 participant senior managers (16 GMs and their immediate assistants) were chosen – representing 4 and 5* in Athens, Thessaloniki, Rhodes and Crete.

Table 4: Case Selection Process

Region	Initial No. of 4 & 5* hotels	Short by Criterion No.1	Short by Criterion No.2	Final No. of Hotels
City Hotels				
Athens	49	26	15	**15**
Thessaloniki	28	20	6	**6**
Resort Hotels				
Crete	250	141	66	66
→ Region criteria narrowed in the area of Chania				**6**

Dodecanese	171	70	56	56
→ Region criteria narrowed in Faliraki & Ixia, Rhodes				**26**

The 16 establishments selected for this research, represent two broad hotel types operating in Greece – city and resort. Basic prerequisite for the participant hotels was to be holders of 4 or 5* official rating that is accredited by the Greek Chamber of Hotels. The ownership status of each hotel (family; local chain; national chain; multinational chain) was also considered. The luxury hotels in the selected geographical regions were then shorted / filtered by using the following two criteria (table 4):

- As a minimum standard the city hotels should provide TV and air conditioning in room and, restaurant and parking facilities. Additionally for resort hotels they should have outdoor swimming pool.

- All participant hotels should have more than 150 rooms. This happened in order to ensure that only medium to big companies would be researched. This aimed to a) compare hotels with similar organisational structure, and b) allow replication in other European countries with similar size and structure hotels.

The case selection process followed in this research was dictated by the structure of the luxury hotel industry in Greece: given its nature and geographical spread (approximately 1,150 establishments all over Greece) a decision was made to limit the destinations in the most representative and popular places for city and resort hotels respectively.

A three-part tool followed by a cover letter explaining the aim of the interview was used, in order to serve the needs of the research. The first part examined demographic data of the company and the participant; the second part was a 14 question semi-structured in-depth interview; and the third part was the Personal Competencies Framework (PCF) Questionnaire, originally developed by Dulewicz and Herbert (1991/99). Additional qualitative data sources derived from observation (field notes) and company documents.

A major methodological concern for this study was to produce valid and reliable outcomes. A research protocol was used as recommended by Yin (1994, 2003). The case study protocol contains procedures and general rules that should be followed in using the research instrument/s and is considered essential in a multiple-case study (Yin, ibid.). It was created prior to the data collection phase. In addition, during the data collection tests for the quality of research were employed (Construct and External Validity, Reliability); these tests were followed by the use of triangulation methods (Data/Theory/ Methodological Triangulation).

3.2. Discussion of the findings

The research findings showed that a Greek GM in his/her career is likely to work in both city and resort hotels, so the roles performed and competencies framework adopted are changing according to the case (best fit approach) in order to cope with the position's demands. 'What, why and how' each GM does in the work context, is the result of a complex process which integrates elements such as personality, roles, competencies, organisational and national context (culture). Since the City hotel type was introduced in Greece quite recently – just prior to the Olympic Games in 2004 – it would be unfair to build the profile of the GMs in Greece based on the "City-Resort" hotel distinction. This research has identified significant differences and similarities of the hotel GMs, when comparisons are based on the ownership status of the company they work.

The first category identified is the Family/Local Hotel Chain, which represent the vast majority of the Greek 4 and 5* hotels (Hellenic Chamber of Hotels, 2007). This is a typical SMTE (small-medium tourism enterprise) owned and essentially co-managed by the leader of the family surrounded by relatives in various positions (In-Group Collectivism). The following category, the National Greek Hotel Chain is a former family business – led very often by a charismatic founder – which expanded gradually its operations nation wide. This type of hotel has adapted to a certain degree the organisational structure and standards of a multinational hotel chain; there is still however moderate involvement of the owner (or his family) to the management of the company. Finally, the Multinational Hotel Chain is a foreign brand name, franchised in most of the cases by a Greek businessman. There are only a few cases that the management of the company belongs to the parent company. This type follows the organisation, structure and standards dictated by the parent company; there are however some variations / deviations due to the Greek socio-cultural context. For example, the standard operating procedures are adapted to the local working patterns and legislation.

The GMs working in a 4* or 5* family/local chain hotel are male between 55-65 years old, speaking in average two foreign languages and have at least a hospitality first degree. Employers in this category are in favour of the "old school" for two main reasons: they value more the experience, reputation and seniority than qualifications; in addition "near retirement" GMs may cost less in the payroll. The recruitment is conducted mainly through recommendations and "word of mouth", and rarely with internal recruits; the selection process is usually conducted by the owner and in most of the cases is based in subjective criteria (i.e. personal references, reputation and salary). There are limited options for training and development in this type of hotel, and very often is up to the GMs' discretion to recommend which programme to attend. In most of the cases, there is no time allocated for training and development activities, in the GMs daily schedule. The job roles performed by the GMs are focused in what Mintzberg (1973) describes as "figurehead", the person who is there to inspire and lead the staff; they also find the time to communicate with customers and listen carefully to their views. The communication competencies cluster is perceived as the most valuable for successful operations and management. GMs in family hotels have an informal performance evaluation – in most of the cases conducted by the hotel owner – based primarily in the financial performance, and secondarily the levels of customer satisfaction and quality. This type of GMs put great emphasis in networking, and they work very hard to build a good reputation in the marker. Their overall relations with the owners can be described as "tolerable" since the GMs are often faced with unrealistic demands on behalf of the owners. On the other hand there are average performers (GMs) who promise more than can deliver; these individuals sooner or later are marginalised and are forced to relocate or work in lower hotel categories (usually 3* hotels). Overall, the level of the owner's involvement (and his family) in the GMs' work in most of the cases is high. The Greek context is dominant here, with the "in-group collectivism" dimension to dictate the relationships between the owner, the GM and their subordinates.

The second type of GMs, those working in Greek National Hotel Chains are males between 45-55 years old, speaking in average two foreign languages and have very good educational background including a hospitality first degree and postgraduate studies. Their professional background shows experience from the 'primary' departments of a medium/big size hotel (Food and Beverage, Front Office - Reservations); in addition, sales and contracting background is a prerequisite for this type of GMs. Recruitment is conducted through personal recommendations or internal candidates with experience in various hotels of the chain; 'head hunters' are rarely used for high profile candidates. Since the recruitment process does not involve a lot of candidates, two or three selection interviews take place with senior managers from/in the Head Office; during the final interview the owner is also present. Throughout the year there are moderate opportunities for training and development; the GMs are free to choose between in-house or outsourced programmes, in Greece and/or abroad. Their job roles are focused in leadership (employee motivation / inspiration) and entrepreneurship (help business grow). The results orientation competencies cluster is their primary concern, they value however the remaining managerial competencies as integral parts of their competencies framework. This is reflected in their performance evaluation, a formal procedure which takes place one or two times a year depending on the type of the hotel unit (city-resort). The primary targets are mainly financial and the maintenance of quality standards; there is however a reference to the 'performance' of the GMs in areas such as communication, leadership and inter-personal relations. The GMs 'secondary' competencies are evaluated through peer reviews, customer satisfaction questionnaires and 'mystery guest' audits. Although there is intense networking activity within the corporate limits, GMs maintain their contacts outside the company; in addition, their reputation is mostly heard within the corporate limits. The owners – who in most of the cases occupy the position of the managing director or chairman of the board – have a moderate involvement in the GMs' work, mainly at strategic level. There are however cases that intervene in GMs' work when they have personal interest, i.e. 'strongly recommend' the selection of a particular candidate. It is important to note here that the owners know personally all of their GMs, and maintain regular communication. Finally, in this type of business the Greek culture meets the corporate culture: the Greek hotel national chains are structured and managed according to the multinational hotel chain model; the Greek culture is however evident everywhere and it is very often the case that 'favours' and deviations from the standards occur when is about relatives or friends. On the other hand, it can be argued that this type of business has embodied the Greek context characteristics in the best way, so their GMs can use it in a beneficial manner.

Table 5 summarises the findings of this research; the three different profiles identified for Greek luxury hotel GMs are not exclusive and provide a generic context for discussion in this field.

Table 5: The GMs' profile in Greek 4 and 5* hotels

Company Type GMs' Characteristics	Family /Local Chain	Greek National Chain	Multinational Chain
Average Age	55-65	45-55	45-55
Sex	Male	Male	Male
Education	HE Graduates	HE Graduates & Postgraduate Edu.	HE Graduates & Postgraduate Edu.
Professional Background	All Departments (Emphasis in F&B)	All Departments (Emphasis in Sales & Contracting)	All Departments (Emphasis in Sales & Contracting)
Recruitment & Selection	Recommendations	Recommendations & Internally	Head Hunters & Internally
Training & Development	Sporadic – GMs' own discretion	Moderate to High Opportunities	High Opportunities
Job Roles	"Figurehead"	Entrepreneur & Leader	Entrepreneur
Competencies	Emphasis in Communication	Results Orientation	Results Orientation

Performance Evaluation	Informal Annual	Formal Annual (1 or 2 times)	Formal Annual (1 or 2 times)
Role of Networking	High	Moderate outside High inside	Low outside High inside
Role of reputation	High in local /national market	High in national marker	High in regional / international market
Ownership level of involvement	High to Moderate	Moderate	Low
Role of Culture	High	Moderate	Moderate to low

The last type of GMs, are those working in multinational companies. They are middle aged (45-55 years old) enthusiastic males with impeccable educational background. They speak in average two languages - including the parent country's (in case it is not English). Their professional background has a sales and finance orientation, although they understand very well hotel operations. The recruitment is conducted internally or through the use of 'head hunters' who are aiming at high profile recruits. In the case of franchised brands the personal recommendation is also used. The selection process is rigorous and involves at least three interviews. There are many opportunities for training and development in Greece and abroad on a regular basis. The GMs' roles in this type of hotels are focused in entrepreneurship and finance. Their annual performance evaluation is multi-dimensional, lots of emphasis is put however in achieving agreed (financial) targets. This corresponds to their preference in the results-orientation competencies cluster. Networking is very important within the corporate limits; outside these limits the GMs maintain only those contacts necessary to 'do the job'. Their reputation is synonymous with hard work and what is actually on their resume. The Greek culture is something that they cannot ignore –especially in the case of foreigners – the corporate culture however is this, which determines their behaviour. It should be noted here that there are less than 10 foreign GMs in 4 and 5* hotels in Greece (most of them in Athens); they are not represented in this study because it was not possible (politely rejected) to reach them. The above profile refers to Greek nationals working in Multinational hotel chains. The fact that a so small number of foreign nationals work as luxury hotel GMs in Greece may lead in the following arguments: first that there are very good Greek GMs who satisfy the high standards of the multinational hotel chains; and second that the Greek context is presenting difficulties that foreign nationals cannot cope with.

3.3. Conceptual Framework

The findings of this research have led to the construction of the following conceptual framework (figure 2) regarding the GM's competencies and roles profile in Greek 4 and 5* hotels. Everything occurs under the influence of a strong national culture which is present within and outside the GMs' working environment. Throughout their careers, GMs initially acquire their competencies by attending higher education courses (in Greece and abroad); this shapes the first generic managerial competencies framework which is consistent with the western conceptions of management competencies (behavioural approach). This generic competencies framework is enhanced and developed through their professional development, in which GMs are shaping their own (personal) managerial profile and competencies framework. They are also 'exposed' to the community of the GMs through their networking efforts and contacts that are developed gradually as they change jobs and positions. Thus, throughout their career, GMs shape a generic competencies framework that is 'enhanced' by a new cluster which refers to the Greek context.

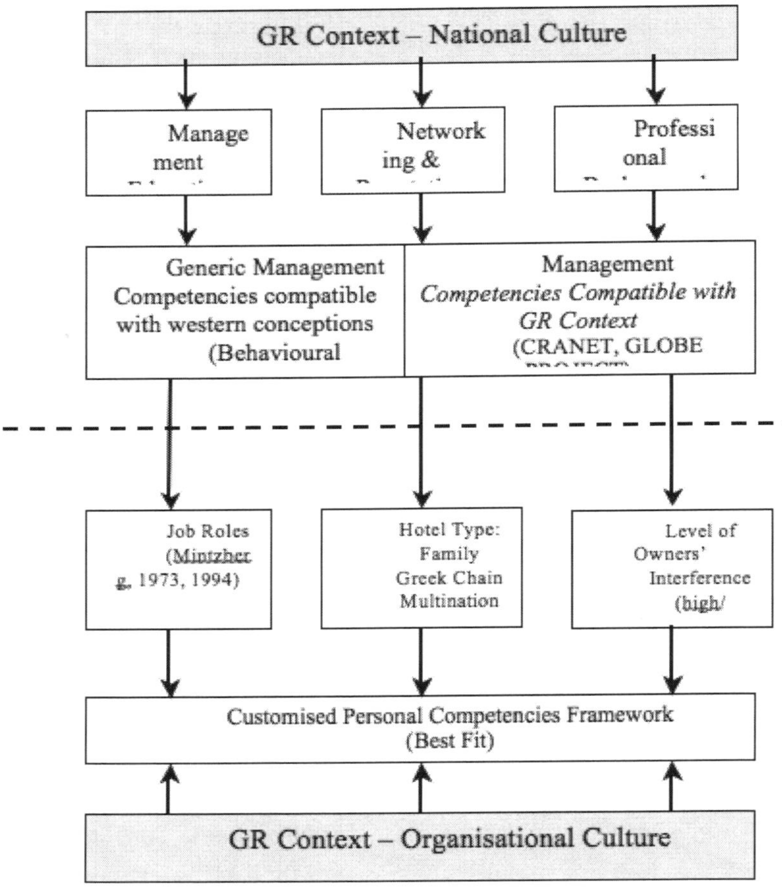

Figure 2: GMs' Competencies Mapping in Greek Luxury Hotels

In their current jobs, GMs are performing their roles according to the competencies framework that has already been shaped throughout their careers; they are influenced however by the ownership status (family vs. chain hotels) and the level of the owners' interference. This situation calls for adaptation to the current position demands, thus the personal competencies framework is adjusted accordingly ('best fit' approach). Again the cultural / contextual factors are present and influence the GMs' roles and competences in their workplaces.

4. CONCLUSION

This research paper has discussed the effects of the Greek cultural context, in the roles and competencies of the GMs' working in 4 and 5* hotels. Previous research related to management and culture in Greece (CRANET survey; GLOBE project), managerial work (Mintzberg, 1973, 1994), and managerial competencies (the 'behavioural' approach), have provided the theoretical framework. The findings of this research identified three different types/profiles of luxury hotel GMs, according to the ownership status of the hotel (family/local hotel chain; national hotel chain; multi-national hotel chain); each one is affected to a certain degree by the strong Greek national culture and this is reflected in their generic profiles. Since research in this topic is still in early stages, the above findings can provide a basis for future research in Greece and/or other European countries.

REFERENCES

1. Anthony P.D. (1986) The Foundation of Management, London: Tavistock.

2. Bartlett C. A. & Ghoshal S. (1989) Managing across boarders, London: Hutchinson.

3. Baum T. (1991) Comparing Expectations of Management Trainees, The Cornell Hotel and Restaurant Administration Quarterly 32(2), 79-84.

4. Baum T. (2002) Skills and Training for the hospitality sector: a review of issues, Journal of Vocational Education and Training 54(3), 343-363.

5. Bourantas D. & Papadakis V. (1996) Greek Management: Diagnosis and prognosis, Int. Studies of Management and Organization 26 (3), 13-25.

6. Boyatzis R. (1982) The competent manager: a model for effective performance, New York: John Willey & Sons.

7. Brewster C. (1994) European HRM: Reflection of, or Challenge to, the American Concept. In P.S. Kirkbride (Ed.) Human Resource Management in Europe: Perspectives for the 1990s, London: Routledge, 56-89.

8. Brophy M. & Kiely T. (2002) Competencies: A new sector, Journal of European Industrial Training 26(2-4), 165-176.

9. Buhalis D. (2001) Tourism in Greece: Strategic Analysis and Challenges, Current Issues in Tourism 4(5), 440-480.

10. Burns T. (1957) Management in Action, Operational Research Quarterly 8 (2), 45-60.

11. Carlson S. (1951) Executive Behaviour, Stockholm: Strömbergs.

12. Carroll S. & Gillen D. (1987) Are the Classical Management Functions Useful in Describing Managerial Work? Academy of Management Review 12(1), 38-51.

13. Cheetham G. & Chivers G. (1996) Towards a holistic model of professional competence, Journal of European Industrial Training 20(5), 20-31.

14. Cheetham G. & Chivers G. (1998) The reflective (and competent) practitioner: a model of professional competence which seeks to

harmonise the reflective practitioner and competence-based approaches, Journal of European Industrial Training 22(7), 267-276.

15. Cheng M., Dainty A. & Moore D. (2003) The differing faces of managerial competency in Britain and America, Journal of Management Development 22(6), 527-537.

16. Child J. (1969) British Management Thought, London: Allen & Unwin.

17. Christou E. & Eaton J. (2000) Management Competencies for Graduate Trainees, Annals of Tourism Research 24(4), 1058-1106.

18. Chung-Herrera B., Enz C. & Lankau M. (2003) Grooming Future Hospitality Leaders: A Competencies Model, The Cornell Hotel and Restaurant Administration Quarterly 44(3), 17-25.

19. Cockerill A.P. (1989) The kind of competence for rapid change, Personnel Management 21(9), 52-56.

20. Dann D. (1990) The nature of managerial work in the hospitality industry, International Journal of Hospitality Management 9(4), 319-334.

21. Dulewicz V. & Herbert P. (1999) Predicting advancement to senior management from competences and personality data: a 7-year follow up study, British Journal of Management 10, 13-22.

22. Eder R.W. & Umbreit W.T. (1989) Measures of managerial effectiveness in the Hotel Industry, Hospitality Education Research Journal 13(1), 333-341.

23. Eurostat (2005), Eurostat Yearbook, Brussels, European Community.

24. Ferguson D. H. & Berger F. (1984) Restaurant Managers: what do they really do? The Cornell Hotel and Restaurant Administration Quarterly 25(1), 27-38.

25. Fondas N. & Stewart R. (1994) Enactment in managerial jobs: a role analysis, Journal of Management Studies 31(1), 83-103.

26. Gore J. (1995) Hotel managers decision making: can psychology help? International Journal of Contemporary Hospitality Management 21(3), 19-23.

27. Hales C. (1999) Why do managers do what they do? Reconciling evidence and theory in accounts of managerial work, British Journal of Management 10, 335-350.

28. Hellenic Chamber of Hotels (2007) Annual Industry Review, Athens (in Greek).

29. House R., Javidan M., Hanges P. & Dorfman P. (2002) Understanding Cultures and implicit leadership theories across the globe: an introduction to project GLOBE, Journal of World Business 37, 3-10.

30. Iversen O. (2000) Managing people towards a Multicultural workforce: Part A – Managerial Competencies, NIPA, Lysaker, Norway.

31. Javidan M.J. & House R.J. (2001) Cultural Acumen for the Global Manager: Lessons from Globe Project, Organizational Dynamics 29(4), 289-305.

32. Kanellopoulos Ch. (1990) Personnel management and personnel managers in Greece, Athens, Greek Productivity Centre (in Greek).

33. Kay C. & Moncarz E. (2004) Knowledge, Skills, and Abilities for Lodging Management Success, The Cornell Hotel and Restaurant Administration Quarterly 45(3), 285-298.

34. Kay C. & Russette J. (2000) Hospitality Management Competencies, The Cornell Hotel and Restaurant Administration Quarterly 41(4), 52-63.

35. Kotter J. (1982) The General Managers, New York: The Free Press.

36. Ley D.A. (1980) The effective GM: leader or entrepreneur? The Cornell Hotel and Restaurant Administration Quarterly 21(3), 66-67.

37. Lockwood A. (1993) European Management Skills in the Hospitality Industry: based on the final report of the HCIMA's European research 1992-4, London, Hotel & Catering International Management Association.

38. Lucia A.D. & Lepsinger R. (1999) The Art and Science of Competency Models, San Francisco: Jossey-Bass.

39. Luthans F., Rosenkrantz S. & Hennessey H. (1985) What do successful managers Really Do? An Observation Study of Managerial Activities, The Journal of Applied Behavioural Science 21(3), 255-270.

40. Martin N. (1956) Differential decisions in the management of industrial plant, Journal of Business 29(4), 249-260.

41. McClelland, D. C. (1973) Testing for competence rather than for intelligence, American Psychologist 28, 1-14.

42. MCI (1991) Personal Competence Project: Summary report, London, Management Charter Initiative

43. Mintzberg H. (1973) The nature of Managerial Work, New York: Harper & Row.

44. Mintzberg H. (1994) Rounding out the Manager's Job, Sloan Management Review Fall, 11-26.

45. Mount D. J. & Bartlett A. L. (1999) The managerial role assessment survey: Design and test of an instrument measuring Mintzberg's roles among hotel managers, Journal of Hospitality and Tourism Research 23(2), 160-175.

46. Mullins L. & Davies I. (1991) What makes for an effective hotel manager? International Journal of Contemporary Hospitality Management 3(1), 22-5.

47. Myloni B., Harzing A.W. & Mirza H. (2004) Human Resource Management in Greece: Have the colours of Culture faded away? International Journal of Cross Cultural Management 4(1), 59-76.

48. Nailon P. (1968) A study of management activity in units of a hotel group, Guildford: University of Surrey MPhil thesis.

49. Nebel E.C. & Ghei A. (1993) A conceptual framework of the hotel general manager's job, Hospitality Research Journal 16(3), 27-37.

50. Papalexandris N. & Chalikias J. (2002) Changes in training, performance management and communication issues among Greek firms in the 1990s: intercountry and intracountry comparisons, Journal of European Industrial Training 26(7), 342-352.

51. Papalexandris N., Chalikias J. & Panayotopoulou L. (2002) Societal Culture and Human Resource Management: Exploring the mutual interaction in Greece, paper presented at the 2nd International Conference Human Resource Management in Europe: Trends and Challenges, Athens, Greece, 17-19 October 2002.

52. Peacock M. (1995) 'A job well done': hospitality managers and success, International Journal of Contemporary Hospitality Management 7(2/3), 48-51.

53. Pickworth J.R. (1982) Managerial jobs in chain organizations, The Cornell Hotel and Restaurant Administration Quarterly May, 30-33.

54. Rees B. (2003) The Construction of Management: Gender Issues and Competence Techniques in Modern Organizations, London: Edward Elgar.

55. Rees B. & Garnsey E. (2003) Analysing Competence: Gender and Identity at Work, Gender, Work and Organization, 10(5), 551-578.

56. Ruth D. (2006) Frameworks of managerial competence: limits, problems and suggestions, Journal of European Industrial Training 30(3), 206 -226.

57. Sayles L. R. (1964) Managerial Behaviour, New York: McGraw-Hill.

58. Schroder H.M. (1989) Managerial Competence: The key to excellence, Iowa: Kendal / Hunt.

59. SETE (2003) Tourism and Employment Report, Athens, Greek Tourism Enterprises Association (in Greek).

60. Sparrow P. & Hiltrop J. (1997) Redefining the field of European Human Resource Management: A battle between national mindsets and forces of business transition? Human Resource Management 36(2), 201-219.

61. pencer L. & Spencer S. (1993) Competence at work: Models for Superior Performance, New York: John Wiley & Sons.

62. Stewart R. (1967) Managers and their jobs, London: McMillan.

63. Stewart R. (1976) Contrast in management, London: McGraw-Hill.

64. Stewart R., Smith P., Blake J. & Wingate P. (1980) The District Administrator in the National Health Service, London, King Edward's Hospital Fund.

65. Tas R. (1988) Teaching Future Managers, The Cornell Hotel and Restaurant Administration Quarterly 29(2), 41-43.

66. Townley B. (1999) Nietzsche, competencies and ubermensch: reflections on human and inhuman resource management, Organization 6(2), 285–305.

67. Umbreit T. (1993) Essential Skills: What graduates need to succeed, Hosteur, 10-12.

68. Whitley R. (1989) On the Nature of Managerial Tasks and Skills: Their Distinguishing Characteristics and Organisation, Journal of Management Studies 26, 209–224.

69. Woodruffe C. (1993) What is meant by a competency? Leadership & Organizational Development Journal 14(1), 29-36.

70. Worsfold P. (1989) Leadership and managerial effectiveness in the hospitality industry, International Journal of Hospitality Management 8(1), 51-62.

71. www.wttc.com (World Travel & Tourism Council) Country Report: Greece (2005), accessed: 11/10/2007.

72. Yin R.K. (1994) Case study research – design and methods, applied social research methods series, (2nd ed.), Beverly Hills: Sage.

73. Yin R.K. (2003) Case Study Research Design and Methods, (3rd ed.), London: Sage.

Chapter 6

Food store owners' and managers' perspectives on the food environment: an exploratory mixed-methods study

Clarence C Gravlee[1], P Qasimah Boston[2,3,4], M Miaisha Mitchell[4,5], Alan F Schultz[6], and Connie Betterley[7]

[1]Department of Anthropology, University of Florida, 1112 Turlington Hall, PO Box 117305, Gainesville, FL 32611-7305, USA.
[2]Department of Children & Families, Substance Abuse and Mental Health Program Office, 1317 Winewood Blvd, Bldg 6, Rm 256, Tallahassee, FL 32399, USA.
[3]Project FOOD Now, Inc., 635 E. College Avenue, Tallahassee, FL 32301, USA.
[4]Tallahassee Food Network, Inc., P.O. Box 365, Tallahassee, FL 32302, USA.
[5]Greater Frenchtown Revitalization Council, 812 Goodbread Lane, Tallahassee, FL 32303, USA.
[6]Department of Anthropology, Baylor University, One Bear Place #97173, Waco, TX 76798-7173, USA.
[7]Leon County Health Department, 2965 Municipal Way, Tallahassee, Florida 32304, USA.

ABSTRACT

Background

Neighborhood characteristics such as poverty and racial composition are associated with inequalities in access to food stores and in the risk of obesity, but the pathways between food environments and health are not well understood. This article extends research on consumer food environments by examining the perspectives of food-store owners and managers.

Methods

We conducted semistructured, open-ended interviews with managers and owners of 20 food stores in low-income, predominantly African American neighborhoods in Tallahassee, Florida (USA). The interviews were designed to elicit store managers' and owners' views about healthy foods, the local food environment, and the challenges and opportunities they face in creating access to healthy foods. We elicited perceptions of what constitutes "healthy foods" using two free-list questions. The study was designed and implemented in accord with principles of community-based participatory research.

Results

Store owners' and managers' conceptions of "healthy foods" overlapped with public health messages, but (a) agreement about which foods are healthy was not widespread and (b) some retailers perceived processed foods such as snack bars and sugar-sweetened juice drinks as healthy. In semistructured interviews, store owners and managers linked the consumer food environment to factors across multiple levels of analysis, including: business practices such as the priority of making sales and the delocalization of decision-making, macroeconomic factors such as poverty and the cost of healthier foods, individual and family-level factors related to parenting and time constraints, and community-level factors such as crime and decline of social cohesion.

Conclusions: Our results link food stores to multilevel, ecological models of the food environment. Efforts to reshape the consumer food environment require attention to factors across multiple levels of analysis, including local conceptions of "healthy foods", the business priority of making sales, and policies and practices that favor the delocalization of decision making and constrain access to healthful foods.

KEYWORDS

Food stores, Food environment, Health inequalities, Formative research

1. BACKGROUND

In May 2010, the White House Task Force on Childhood Obesity identified food environments as a driver of childhood obesity and recommended a $400 million initiative to increase access to healthy foods in low-income communities in the United States [1]. This initiative and related efforts draw on evidence that neighborhood characteristics such as poverty and racial composition are associated with inequalities in access to food stores [2–8] and in the risk of obesity [9, 10]. The assumption is that unequal access to food stores constrains dietary choice and contributes to social inequalities in the risk of obesity and diet-related disease [11].

Research in this area has been guided by the distinction between community and consumer food environments [12]. The *community food environment* refers to the spatial distribution of food sources, including the location, density, and type of food stores. Evidence linking community food environments to health, so far, is mixed [13–15]. Some studies report associations with dietary intake [16–18], but others show no association [19–22] or associations in an unexpected direction [23, 24]. The inconsistency of results is partly attributable to differences in measurement [25], but it also likely reflects the distal relationship between community food environments and dietary behavior: The effect of community food environments on dietary intake depends on many intervening factors at the local level, including the quality of food stores and products available in them, local marketing and promotion strategies, the quality of service in stores, and residents' perceptions of local food stores [26].

Research on the consumer food environment addresses this limitation. The *consumer food environment* refers to what consumers encounter in and around food outlets, including the placement, cost, quality, and promotion of specific foods. Previous observational research suggests that consumer food environments are linked to dietary intake in multiple settings. In a study of 12 communities in California and Hawaii, for example, the availability of low-fat and high-fiber foods in local retail outlets was associated with reported healthfulness of individual diets [27]. Among African American women in eastside Detroit, more positive perceptions of the selection and quality of fresh produce at local retail outlets was associated with higher fruit and vegetable intake, independent of store type and location and of individual-level characteristics such as income and education [28]. And in New Orleans, vegetable intake of local residents was positively associated with the amount of shelf space allotted to vegetables in food stores located within a city block of one's residence [29].

These findings have led some researchers to target food stores in obesity prevention interventions [30]. A recent systematic review identified 16 interventions in small food stores over the 20 year period of 1990–2010 [31]. A common goal across all the trials was to increase the stocking of healthy foods, especially produce, low-fat dairy, and whole-grain breads; some interventions also tried to reduce the availability of unhealthy foods. Other common strategies included in-store signage to promote healthy foods, cooking demonstrations or taste tests to increase familiarity with healthy foods, stakeholder workshops to develop or refine interventions, and community meetings designed to engage store owners and community members. Results indicate that store-based interventions have modest but positive impacts on the availability, sales, and consumer knowledge of healthy foods [31].

Despite such promising results, however, we know relatively little about store owners' and managers' perspectives and experiences of factors that impact the consumer food environment and may affect the efficacy or sustainability of future interventions. Of the 16 trials identified by Gittelsohn and colleagues [31], for example, only three included interviews with food store retailers. Two of these studies [32, 33] reported process evaluations to assess store employees' experiences of the interventions; one [34] reported formative research that led to a successful corner-store intervention. In all three studies, store owners were generally willing to participate in store-based interventions and helped identify potential barriers to increasing sales of healthy foods, including consumer demand and store infrastructure (e.g., refrigeration). Yet it is unclear whether lessons learned in these studies sufficiently inform store-based interventions in other settings.

This paper extends the evidence base for future food-store interventions by examining the perspectives of store owners and managers in Tallahassee, Florida. We conducted an exploratory, mixed-methods study to understand the business practices and contextual factors that influence decisions about what and how to sell the products available in food stores in Tallahassee. We also elicited store owners' and managers' perceptions of the food environment, including their expectations for how, if at all, they could increase access to healthy foods. In this sense, our study contributes to understanding how the consumer food environment comes to be and how best to intervene. We also argue that our findings enhance multilevel, ecological models of the food environment and challenge us to confront broader social structures that drive social inequalities in health.

2. METHOD

2.1 Research setting

Tallahassee is a small city in the panhandle region of Florida (USA), with an estimated population of 181,376. In the 2010 US Census, 57.4% of the population self-identified as White, 35.0% as Black, and 3.7% as Asian. Approximately 6% self-identified as Hispanic or Latino. As the state capital and home to two major universities, Tallahassee's largest employment sectors include education, public administration, health care, and the arts and recreation. During 2006–2010, an estimated 28.5% of residents were living below the federal poverty level.

Previous researchers have documented spatial inequalities in the community and consumer food environments of Tallahassee and broader Leon County. Rigby et al. [35] examined the distribution of food stores in Leon County by neighborhood characteristics, including poverty and racial composition. They found that small grocery stores accepting Supplemental Nutritional Assistance Program (SNAP) benefits were more common in predominantly Black neighborhoods, but these neighborhoods had no supermarkets. Leone et al. [36] showed that store type and neighborhood characteristics are associated with the availability of healthy foods within stores in Leon County. In particular, (a) supermarkets offered the lowest prices for fresh fruits and vegetables, low-fat milk, and whole-wheat bread, and (b) stores in high-income neighborhoods devoted more shelf space to healthy foods than did stores in low-income neighborhoods.

We explored the sociocultural processes driving these inequalities by eliciting the perspectives of food store owners and managers in the areas of highest need. Our work emerged from a community-academic partnership known as the Health Equity Alliance of Tallahassee (HEAT), which is grounded in principles of community-based participatory research [37]. Community partners and local policy makers directed our focus to two areas defined by the City of Tallahassee's Community Redevelopment Agency (CRA) in 1998. The CRA includes two discrete redevelopment areas that border downtown: Greater Frenchtown and Southside (Figure 1). Both areas include predominantly African American neighborhoods with relatively high levels of poverty but also rich histories of identity, community, and civil rights activism [38, 39].

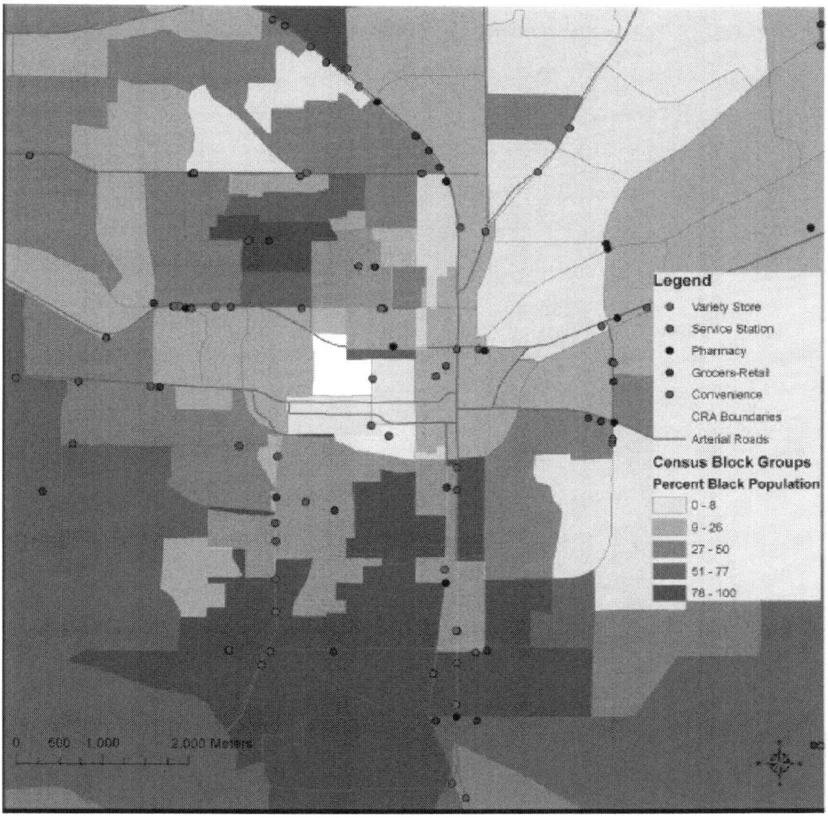

Figure 1 Racial composition and the location of food stores in the vicinity of the Community Redevelopment Area (CRA), Tallahassee, FL, 2010.

2.2 Participants

The Leon County Health Department provided a complete list of 59 food stores in the vicinity of the CRA areas. We consulted with community-based organizations and local residents to ensure that the list was accurate and complete. From this list, we selected a purposive sample of 25 food stores to represent variation in store type and neighborhood characteristics (poverty and racial composition). Sample size was based on theory [40] and empirical research [41] about the minimum sample needed to capture variation in cultural knowledge and experience in a particular domain. The sample included two supermarkets, five grocery stores, eight convenience stores, four gasoline stations, and one pharmacy. We recruited one manager (n = 16) or owner (n = 4) from each store that agreed to participate. Potential participants in five stores refused to participate. Some managers indicated that corporate policy prohibited them from participating in research; others may have faced time pressures or may have doubted our explanation of why we were doing the study.

2.3 Interview design

We conducted in-store interviews during business hours. This decision allowed interviewers informally to observe the context and flow of business in the store, but it increased the need to conduct interviews efficiently. We chose an open-ended, semistructured interview format to balance the need for efficiency with our exploratory goals of allowing participants to define the pace, content, and flow of the interview [42].

We drew on the local expertise of community partners and the experience of other communities and researchers in designing food-store interviews. The purpose was to give store owners and managers an opportunity to teach us about their perspectives and experiences regarding their business practices and the food environment. Questions were open-ended, and interviewers were encouraged to follow new leads introduced by participants. We included a free-list question [43, 44] in which participants were asked to list as many "healthy foods" as they could. The purpose of this technique was to elicit store participants' perceptions of what constitutes healthy eating, so that subsequent intervention messages could be crafted in terms that are meaningful to store owners and managers.

The interview guide reflects input from university researchers (CCG), community partners (MMM, CB), and local high school and university students who conducted interviews in collaboration with the project coordinator (PQB) as part of a youth leadership development program. The University of Florida Institutional Review Board approved the study protocol (#2009-U-1148); we obtained written informed consent from all participants. Interviews were completed between February and June 2010.

2.4 Analysis

Interviewers audio recorded 14 of the 20 interviews, and we made verbatim transcripts following standardized guidelines [45]. For six participants who did not consent to audio recording, we relied on notes that interviewers took during the interview and expanded immediately afterwards. We used the notes for context and interpretation; the data presented here come from verbatim transcripts of audio-recorded interviews.

We imported free-list data into ANTRHOPAC software (Analytic Technologies, Lexington, KY, USA) to calculate the frequency, average rank, and salience of the "healthy foods" that participants listed [46]. These results help to identify types of foods that store owners and managers consider to be healthy.

We imported interview transcripts and notes into MAXQDA 11 software (VERBI GmbH, Berlin, Germany), which facilitated both word- and code-based approaches to the analysis [47]. First, we examined word frequencies for the whole sample and separately by store type to identify potential themes that we later examined through close reading of the text [48]. Second, we read the texts to identify themes, or underlying dimensions of meaning, with a focus on repetition, similarities and differences within and between texts, and linguistic connectors (e.g., "because", "then", laughter) [49]. We identified themes inductively, based on patterns in the data, in keeping with our exploratory aims. The first author (CCG) developed a codebook to standardize definitions of themes and specify the conditions under which segments of text should be coded as instances of a theme [50]. He did the primary coding, with other team members contributing to interpretation and synthesis.

3. RESULTS

3.1 Free lists of healthy foods

Respondents listed 96 items as "healthy foods" (mean number of items per respondent = 8.3, SD = 6.9, range = 2–33). Most items (77%) were listed by only one or a few respondents, as is common in free lists. Table 1shows the 21 items that were listed by more than one respondent. Three patterns are evident.

First, no item was listed by half of the respondents or more. *Vegetables* and *fruit* stand out as the most frequently listed items, but they were mentioned by only 47 and 42 percent of respondents, respectively. All other items were listed by fewer than one-third of respondents. Typically, this pattern indicates low agreement about what belongs in a cultural domain.

Table 1 Items listed by more than one respondent in free list of "healthy foods" (N = 19)

	Item	Percentage	Average rank	Smith's S
1	Vegetables	47	2.7	0.383
2	Fruit	42	4.0	0.274
3	Chicken	32	6.5	0.183
4	Cheese	32	7.0	0.153
5	Juice	26	3.8	0.160
6	Milk	26	4.0	0.182
7	Bread	26	6.4	0.126
8	Meat	21	4.5	0.077
9	Cereal	21	5.0	0.118
10	Fish	21	4.8	0.136
11	Peas	16	12.7	0.080
12	Nuts	16	6.0	0.094
13	Eggs	16	4.7	0.097
14	Soup	16	7.0	0.077
15	Salad	16	5.3	0.103
16	Baked foods	11	3.5	0.071
17	Greens	11	4.0	0.058
18	Oranges	11	12.0	0.034
19	Onions	11	5.0	0.046
20	Apples	11	11.0	0.047
21	Water	11	2.0	0.084

Second, in general, food-store managers and owners listed items that public health nutritionists would also identify as healthy foods. Fresh produce dominates the list, both in the generic categories *vegetables* and *fruit*and in specific items such as *peas, salad, greens, oranges, onions,* and *apples.* The list also includes dairy, low-fat meats (e.g., *chicken* and *fish*), and healthy cooking methods (e.g., baked foods). This pattern suggests opportunities to craft intervention messages using concepts and categories that are meaningful to store owners and managers.

Third, free-lists results raise questions about possible cultural differences between what store owners perceive to be healthy foods and what consumers prefer to eat. There is little evidence, for example, that store owners and managers' perception of healthy foods is tailored to the cultural preferences of customers in the predominantly low-income, African American neighborhoods the stores serve.

Following the free-list question, we asked respondents to list the healthy foods, if any, sold in their store. Table 2 shows the 16 foods identified by more than one respondent. Again, the low frequency of all items indicates relatively little agreement among respondents. On average, respondents identified about half as many healthy foods in the store as they did in the free-list (mean number of items per respondent = 4.3, SD = 2.2, range = 1–8).

In particular, respondents reported that few healthy foods, as defined by their free lists, are available in the stores. Vegetables and fruits dominate the free lists but are not commonly among the "healthy foods" available in the sampled stores. *Bananas* and *apples* were mentioned by multiple respondents, but no fresh vegetables were. Instead, processed foods are prominent in responses about the healthy foods available in stores. Such "healthy foods" include *canned vegetables, fruit cups, juice drinks, cereal,* and a variety of snacks. Perishable items are virtually absent from the list.

According to store managers and owners, the most common "healthy food" in stores is snack bars. The terminology for this category is not uniform but includes *cereal bars, granola bars, Nutri-Grain bars, Power Bars, health bars, energy bars,* and *protein bars.* If these items were collapsed into a single category, it would be the most common item of all (with a frequency of 10, or 50 percent).

Table 2 Items identified by ≥2 store managers and owners as "healthy items" offered in the store (N = 20)

	Item	Percentage	Average rank	Smith's S
1	Juice	35	3.4	0.188
2	Milk	35	3.0	0.213
3	Canned Vegetables	20	2.5	0.156
4	Cereal	15	4.0	0.088
5	Bananas	15	4.7	0.051
6	Orange Juice	15	2.7	0.087
7	Fruit	15	3.3	0.087
8	Chicken	10	1.5	0.093
9	Fruit Cups	10	1.5	0.087
10	Grapefruit Juice	10	4.5	0.051
11	Cereal Bars	10	1.5	0.093
12	Cheese	10	4.0	0.049
13	Apples	10	5.5	0.029
14	Granola Bars	10	1.0	0.100
15	Nutri-Grain Bars	10	3.0	0.073
16	Soup	10	4.5	0.044

3.2 Thematic analysis

We have organized the themes from semistructured interviews into four levels of analysis: (1) store and business factors, (2) economic factors, (3) individual and family-level factors, and (4) community factors. Figure 2 summarizes the most salient themes in each area.

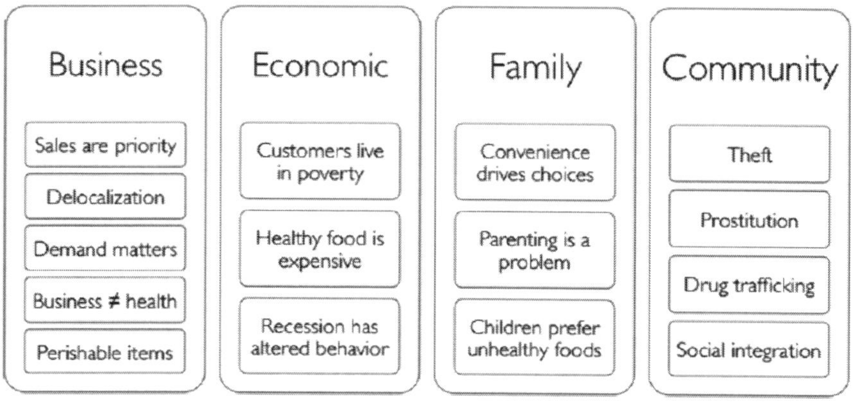

Figure 2 Key themes in semistructured interviews with food-store owners and managers (N = 20).

3.3 Store and business factors

3.3.1 Business priority is making sales

Not surprisingly, the top priority for managers and owners in all store types is to make sales. In the words of one drug store manager, "Sales is what drives it". A local grocery store manager and the manager of a chain convenience store, respectively, offer typical explanations for how they decide what to sell:

Pretty much how fast it sells. Costs, definitely—cost plays a part of it. But...it's pretty much how fast it sells, you know. We're not gonna get it if it don't sell. (SS14)[a]

You know, what they'll [corporate purchasing] do, they'll take a wide range of products. Say, for instance, food. We may get different types of food in here for six months. If that product turns a lot, then it's gonna become a permanent product. If it doesn't turn—we don't get a whole lot of turns on it—then we're not gonna continue to carry it. We'll clearance it out. (QB02)

Another manager emphasized that the store sells "whatever the clientele wants. That's from corporate down, 'cause we're sales-driven, obviously" (QB04).

3.3.2 Delocalization of decision making

The previous respondent's reference to decision-making "from corporate down" highlights the delocalization of decision-making about which products to sell. Local managers generally perceive that they have little influence over the foods available in their stores, referring instead to "corporate", "my corporate manager", or "that's a corporate thing".

That's really a corporate thing. That comes down from corporate, and district has some things—our regional and district office has, obviously, some say in it. But the majority of things come from our corporate office, 'cause that's where our purchasing department is at—our purchasing and marketing department. So, they kind of make a group decision, and they try to pass it along to all the stores, let's say, in Florida. And so most of the stores in Florida are all gonna carry the same thing. It helps with distribution and pricing and things like that. (MR21)

The manager of a chain supermarket offered a similar explanation:

A lot of that comes from corporate. We can, if a customer requests something, we can go through corporate and request, "Hey, we've got a customer..." Late last week [I had a customer] wanting rice drink, the rice milk? And I'm trying to get that for her right now. It's not approved at corporate, but I'm trying to get it for her. (QB04)

Corporate decision-making limits store managers' perceptions of how they could influence access to healthy foods. The manager of a convenience store said: "As the store itself, we have no control over what we sell. So there's nothing I can do on my part to make those changes" (QB06). A drug store manager elaborated:

Really, as an assistant manager, we don't really have much say in what we carry. Yeah, we can put in our two cents, but even at the store manager's level, unfortunately, it still goes back up to corporate. So, yeah, we can tell them, "We need this and we need that," but, unfortunately, they're gonna go off of what's gonna be the best price, what they're gonna be able to keep on the shelf the longest, because they can't keep a lot of fresh foods. It would be nice to see some fresh fruits up at the front register and things like that, 'cause I think they would sell...but I just don't see us, at the store level, having enough influence to make those changes. (MR21)

The regional or national market of chain stores limits the ability to focus on specific communities. As one manager explained, "Companies like this that got 5,000 stores will—maybe only a few hundred might be in a neighborhood like this, so they try to offer a variety of stuff based on where you're located, as a whole". He continued:

Only 20 percent of the country is, is minority, or black, whereas the majority of the country is white, so you're gonna carry more products that's gonna cater to a certain population group than, you know... We don't carry anything that would cater to a Hispanic, say... But in central Florida and south Florida they carry more products based on, you know, food products and hair care products based on that demographic.... So if we can get the companies to just concentrate more on the community itself, rather than as a whole, like a city or country—concentrate more on what the community needs.... (QB02)

The delocalization of decision making is clearest for stores that belong to regional or national chains; owners of independent grocery and convenience stores make their own decisions about what to sell in the store. Even independent stores, however, are constrained by non-local, corporate decisions made by the vendors they rely on:

So, [customers] ask me, "Well, you don't have this one. Could we get that?" I say, "Yeah." ... So we're open to suggestion. And that's what I think, a business works best that way, too. But my vendors don't believe it! They want to do what their corporation tells them. I say, well, I'm in between you and the customer. (QB01)

3.3.3 Consumer demand matters

One way store managers can try to influence corporate decision-making is by relaying customer feedback.

About the only thing I can really do is just continue to try to make, you know, give input back, feedback to the company, you know, through emails or what customers are interested in and may want to see in the store. And we try to do that on a weekly basis. (QB02)

This excerpt points to the importance of consumer demand as it relates to the priority of making sales. The perception of customer demand is most important in independent stores, where owners make their own decisions, but even managers of chain stores express this theme.

Most of the time, if some customer comes in and say that, uh, "You don't have this product", and they would like to have it, we go ahead and buy it. And if we see, if we sell more of it, we keep ordering it. If we don't sell it, then, we buy the one time and next time, when the same customer wants it or somebody else wants it, that's when we order it. (MR22)

Interviewer: What would it take for corporate to want to offer healthier foods in the store?

Demand. Public demand. Customer demands. And the popularity of how it's selling in other areas and stuff. (QB06)

Actually, sometimes you have people that come in, and if they don't see it in the store, they will say, like, "Why y'all don't sell this?" or, "Why you don't sell that?" or "Maybe you should try.... So, what I say to him [the owner]—because they're foreigners—I would say, "OK, this is what our people like", you know? And they get it! (SS15)

Some respondents identified lack of demand for fresh produce as a barrier to offering healthier food. This perspective was especially relevant for non-grocery stores, as the manager of a convenience store explained:

I guess we could probably carry a lot more perishable items as far as fruits and stuff like that, but, like I said, if they don't pretty much come in and ask for it, we don't carry it. And a lot of people are finicky about where they buy their products, you know what I mean? So, depending on the setup of your store and exactly where you're at, you're not just gonna come in and order, I mean, go grocery shopping in this store particularly. I don't think anyone would come in and actually grocery shop. Given the area we're in, I just don't think they're gonna go in and purchase fruits and stuff out of this place here. I just don't think they would. (SS14)

3.3.4 What's good for business is not necessarily good for health

The priority of making sales may clash with public health goals of increasing the availability and consumption of healthy foods. This conflict is evident in some respondents' perception that what is good for business is not necessarily good for health.

The most item I sell, you know, contains a lot of sugar, like the Little Debbies, Honey Buns, which is not good these days because the kids don't get enough exercise and don't get involved in sports and stuff at all. So those are the food that's not good for them. Sometimes they can eat donuts—but not every day. And I've seen some people buying it every day, sometimes two times a day, which is not good. And you know, the good thing is, you know, the good thing is the business. I do business because of that.

Interviewer: Right. It's good for your business.

But, I mean, it's not good for health. (SS18)

One respondent expressed dismay that processed snack foods—"the worst thing I ever, ever, ever"—were "the best sellers" in his convenience store (QB01). Another manager replied with embarrassed laughter when asked what he thought were some of the good things about the choices people who come into his store make: "Some of the good things? They're high profit items for me" (QB06).

3.3.5 Stocking and selling perishable items is a challenge

One of the most commonly identified barriers to offering healthy foods is selling perishable items before they expire. This is a complex issue that involves low perceived customer demand, high risk of taking a loss on unsold items, and inadequate infrastructure for stocking perishable goods.

To begin, store managers and owners express an understanding that the healthiest food choices are perishable items. When asked about healthy foods in the store, a drug store manager replied:

We have some in the cooler. Most of our healthy—like I would consider real healthy stuff—is in the cooler. We have some, like, fruit cups. We're just now carrying a grapefruit cup for in the morning, for breakfast alternatives. We don't carry any fresh veggies or fruits. We used to, but the fruits—we mainly had fruits—but they were kind of going bad, so they had to take those out. (MR21)

Other respondents picked up on the challenge of selling items before they expire:

In another way, I've tried, you know, several other foods that are like, you know, lite. They call it 'lite'. You know, snacks and stuff like that. So I tried to keep that in the store, but it went out of date, so I had to stop from keeping it. (SS18)

The short shelf life of perishable items is particularly challenging for owners of small, independent stores, who risk loss if items do not sell. One owner linked his perception of risk to low demand, expiration dates, and distributors' policy of not accepting returns for expired items:

If I buy it, and it doesn't sell, I have to throw it out. The milk guy doesn't deliver anymore. I buy, myself, all the milk. If not, I have to throw it out or drink it myself. They [distributor] add to my minimums about a hundred, two hundred dollars worth of milk, and there's no return on them either. ... It's demand. The supply is there, but then the supply is time-limited. It's no return. That's what the problem is. I cannot keep all the food they [customers] want because all the healthy food has dates on them.

Interviewer: OK, so the expiration dates are a problem?

Yeah, and the vendor doesn't take it back. For me, you know, losses ought to be shared. Like a profit, we're sharing. Why can't we share the losses too? But they don't. (QB01)

Many stores simply lack the infrastructure to stock perishable items. When asked what it would take to stock more healthy foods, the owner of a small convenience store said:

We'd probably have to get another refrigerator. Definitely a cooler or something to carry the meats in—I mean, given that we're limited in space inside this place. We would definitely have to find some kind of way to fit a cooler into where we can actually carry, I guess, some fruits and maybe organic foods and whatnot. (SS14)

The emphasis on space and refrigeration was especially common in non-grocery stores. A gas station manager was clear about what it would take to stock more healthy foods: "I guess more space to put it. I got a smaller type of store here" (SS06).

3.4 Economic factors

3.4.1 Poverty constrains consumer food environment and healthy eating

Store managers and owners see that cost drives down demand for healthier foods. This problem was magnified during the recession of 2008–2010, but even in more prosperous times, respondents perceived that poverty shapes the types of foods people buy and how frequently they shop:

It's a poor community, so a lot of people are basically eating what they can afford, which is like noodles, Ramen noodles and stuff like that—some canned goods. First of the month, you'll have a customer maybe come in here and buy $83 worth of canned goods and noodles and juices because, basically, that's what they can afford. You know, the lower, the cheaper end products (QB02).

Some stores extend credit to customers who do not have enough money to purchase food. This practice puts stores at risk, if customers default on credit payments:

Some people around here, when they come in to buy something, they don't have enough, so we have to work around that and just tell them to bring it next time or something like that. We do have a credit account for some, a few people, so if they don't have money right away, we can put it in the book for them, and when they have money, they come over here and pay it, so... Sometimes we forget, sometimes they forget, so that's mostly the problem we have right now (MR22).

3.4.2 Healthy food is expensive

The economic challenges are compounded by the perception that healthier foods are expensive. The higher cost of healthy food has two consequences for food stores: (1) owners or purchasing managers make decisions about what to buy based in part on costs to the store; (2) the perceived higher cost of healthier foods lowers demand and makes it more difficult for stores to sell perishable items before their shelf life expires. A small grocery store owner summarized how these issues intersect, when asked about how he decides what to stock:

Can people afford it, first of all—affordability. And is it going to sell. That's mainly the thing. If it does not sell, then I have to think twice. Mainly when I buy stuff for the store, I consider myself as a customer—if I'm going to buy it or not. Then I decide that, yeah, I can afford to buy it, then I know that my customers can afford to buy it. I put myself in their shoes. (QB01)

The manager of a small grocery store contrasted the costs of healthy and unhealthy foods in the store:

See, most everything right now is costlier. Beer is the only thing that's the cheapest. We do have some juice that's $1.99, and a lot of college students around FAMU, they buy them a lot. Well, if it were up to me, I would make sure that the price of food is cheaper than soda and everything. The cheapest thing we have is sweets and candy, and people, sometimes they have five or ten cents, they go ahead and buy a *candy instead of saving that and collecting like a dollar or so, so they can buy some nice meat or some juice or something like that. (MR22)*

Other store owners and managers emphasized that the costs of food can be reckoned both in terms of money and time:

It's cheap. I mean, look at the stuff. You can go and buy a dollar hamburger, and you can't go buy a dollar steak. Or a dollar piece of chicken and then take the time to cook it yourself. So, I think the convenience, and the culture behind that.... (MR21)

3.4.3 Recession changed consumer behavior

Store owners and managers reported that the cost of food became a more significant problem during the recession of 2008–2010. The recession was especially difficult for small, independent stores, where respondents lamented that "money is tighter" (QB01) and "it's tough right now, very tough" (SS05).

It's day-to-day getting so hard, and the economy doesn't help at all. People used to buy $0.99 bag of chips. Now they're looking for $0.25 bag of chips. They don't buy that much. I had to look for alternates, as well. And so many of my vendors, they stopped coming here, because I don't sell much. So, I had to look for alternates who can supply the same kind of stuff, so I had to go and shop around, too. And everybody's got just the same price, you know, high price. (QB01)

Some respondents thought the recession reversed a positive trend of buying more healthy foods. A supermarket manager explained:

From my point of view, the past couple of years, since the economy has gotten to the point it's at, it has actually regressed some—people aren't spending as much money, trying to save money where they can. But prior to that, we were seeing an increase in more organics, more fat-frees, more whole wheats, etc. (QB04)

3.5 Individual and family factors

Store managers and owners also identified individual and family-level factors as determinants of unhealthy behaviors. Some respondents linked these factors to structural constraints (e.g., poverty, household composition), while others emphasized what they perceived as cultural patterns.

3.5.1 Convenience drives food choices

Many respondents perceived that convenience drives food choices. This perception relates to the earlier theme of cost: If people lack money or time to prepare food on their own, they may resort to processed or prepared foods:

Yeah, definitely convenience. I mean, versus you going into the kitchen and preparing a meal, that would save you, what? Taking about thirty minutes to prepare a meal versus you just going to McDonald's and getting it within five minutes. I mean it's definitely convenience. ... It's just a lot easier to go and pick it up versus you going in and preparing it yourself. (SS14)

While some respondents regard time poverty as a factor, others invoked cultural explanations, citing the decline of cooking, the pace of lfe, or a desire (promoted by advertising) for convenience.

Looks like they need frozen, ready food, you know? Yeah, put it in a microwave, finish it up. Lately, the young generation don't like to cook, no. They want to go prepared. (SS05)

Probably the culture we're in is too quick, fast. We're gonna go after fast food before we're gonna go home and cook us a meal, or warm us up some soup. (MR21)

It's quick, convenient, and it's advertised, and that's what they have on their mind that they want. (QB06)

3.5.2 Parenting is a problem

Some store owners and managers see parenting as a source of unhealthy food choices. One respondent labeled the problem as "lazy parents". A more common explanation was that parents are under time or economic pressure and do not have time to cook for themselves or to teach children about nutrition. Whatever the cause, respondents commonly linked children's food preferences and purchasing decisions to values they learned at home:

Because they have—I can't say 'no home training'. Their parents don't push them to eat healthy stuff in the house. I betcha no house has apples, bananas, and oranges, like mine. My baby, I push him to eat stuff like that, even though he say he don't want it. If that's all you got, what you gonna eat as a snack, you know? He say, "Ma, go buy Little Debbie". No, I buy apples, oranges..." (PS25)

3.5.3 Children prefer unhealthy foods

Whereas some respondents attributed children's preference for junk food to poor parenting, others saw it as an inevitable part of childhood. When asked why children buy junk food, for example, one convenience store manager replied, "It's just kids being kids, you know?" (QB02). A supermarket manager agreed and related this observation to his own (presumably well parented) children. In reply to the question of why children buy junk food, he said:

As far as why they pick 'em? Because they're not healthy. [laughter] I've got two little girls. They wanna eat junk. [laughter] Like the Capri-Sun?

That's a big thing. And the Kool-Aid Jammers—among the kids, we sell a lot of these. (QB04)

Not all respondents explained why they thought children would prefer unhealthy food over healthier options, but one grocery store manager suggested that it could be an act of rebellion: "I guess they wanna be bad". But then he noted: "Although, when they're good, they get it as a reward. [laughter]. I don't know" (SS16).

As a component of consumer demand, children's food preferences relate to the core priority of making sales. Thus, as one grocery store manager acknowledged with uncomfortable laughter, children's food preferences influence marketing decisions:

Definitely the snacks. I would say the hard candy, the chocolates, ice cream, the chips. Anything that's not healthy, that's what they come here for. [laughter]

Interviewer: So why do you think they buy those foods?

I think that's just something that started early, at an early age, where they're giving the kids sweets and whatever. They get used to it. It's almost as a treat, you know what I mean? You almost rewardin' 'em with something, so they get used to it, so that's what they're gonna come in and ask for. And that's pretty much all they come in and get. That's mainly why we carry what we carry. (SS14)

3.6 Community factors

Store owners and managers identified community-level factors as among the most important challenges facing their business. These factors include theft, loitering, drug trafficking, prostitution, and other criminal behavior near the store. Some respondents also identified a lack of community cohesion as a challenge to their business.

One supermarket store manager said the biggest challenge is "keeping everything at the front cleaned up" (QB06). Other respondents identified illicit activity such as theft, prostitution, and drug trafficking as key challenges:

Our biggest challenge is shoplifting and solicitation outside the store. It's kind of an ongoing problem. You know, we've even had prostitution in the parking lot and just, you know, a lot of people out begging. So we get a lot of interaction with the customers, telling us that we got somebody out in the parking lot begging. So we deal with the police a lot, where we have to get them out here to kind of help us handle that, plus the shoplifting, so that's our biggest issue. (QB02)

Dealing with these challenges increases the cost of running the business because it requires having more staff: "I have to make sure I have enough people on coverage to make sure that we can keep it clean, as well as customer service" (QB06). These costs are then passed on to consumers:

Interviewer: What are the biggest challenges you face in keeping your business going, and how to you deal with those challenges?

Theft. I mean, we put 'no trespassing' warnings, we send 'em to jail, we try to raise prices to back up our inventory, when it falls short. That's how we make up our inventory. (PS25)

Theft, prostitution, and drug trafficking pose a dual challenge to store owners and managers because such activities not only drive up costs but also may drive away customers. A manager of a small grocery store explained that the biggest challenges to her store are "outside activities":

The young people hang around the store. No matter how much we call the police and ask them to pass out trespassing warnings, they still come back. Traffic is kind of congested, too, certain times of day, because of the activities outside.

Interviewer: So do you mean as far as—is it like juveniles hanging out or drug dealing?

It's juveniles, grown people, drug dealers. Just, man, keeping it real, yes.

Interviewer: You think those prohibit certain people from shopping here?

Yes it does. It does. (SS15)

This store manager explained that she thinks the store has a role to play in organizing people in the neighborhood to reclaim ownership of the community. Here is her explanation of how to deal with the "outside activities":

And with me asking other people to get involved in the community—let's take our community back. I think that calling the police and other people getting involved is helping us out a little bit, too. ... I've spoken to a lady named [name]. She says she's a homeowner association president. And I've gotten her involved in this also, and we talk constantly about it. We're supposed to be getting a meeting together because the homes in the community won't sell. The property value is going down. And we just need to get—we, the people, need to get more involved. Even the church, the ministers over here need to get involved. (SS15)

As she continued, this store manager connected community redevelopment back to the consumer food environment:

You know, especially with my older customers. I have a lot of seniors, see, and that's one of the main reasons, too, why I want the community to get involved. I want the senior citizens to be able to come back to their roots—you know what I'm saying?—without feeling threatened by the people that's hanging out front and, you know, feeling uncomfortable with coming in and spending their money, thinking that they're gonna get robbed. I'm not saying that those guys would do that...but, you know, some people just don't have that feeling, that sense of feeling... (SS15)

4. DISCUSSION AND CONCLUSIONS

The purpose of this exploratory study was to understand the business practices and contextual factors that influence decisions about what and how to sell the products available in food stores in Tallahassee, FL. We presented systematic evidence regarding how food-store owners and managers conceptualize "healthy foods" and identified factors at multiple levels of analysis that shape their view of the food environment and their role in providing access to fresh, nutritious foods. The findings enhance multilevel, ecological models of the food environment and imply that reducing social inequalities in access to healthful foods will require interventions across the spectrum of prevention [51].

Our work contributes to a growing literature on the role of food stores in obesity prevention [31]. We identified six previous studies involving open-ended interviews with food-store owners [32, 52–56]. Our results confirm several key findings from this literature, including the priority of making sales, store owners' perception of low customer demand for fresh produce, the challenge of selling perishable foods, and lack of space or infrastructure to stock fresh foods. Store-based interventions designed to address these barriers generally have shown modest, positive impacts on availability, sales, and consumption of fresh foods [31,57], and at least one trial reported an impact on body mass index [58].

Our study builds on this work in three ways. First, we present systematic evidence regarding how store owners and managers conceptualize "healthy foods". This contribution is significant because it could help to frame future store-based interventions in terms that are meaningful to food retailers. In particular, our findings suggest that store owners' and managers' preconceptions about healthy foods overlap with public health messages, but that (a) agreement about which foods are healthy is not widespread, (b) store owners' perceptions of healthy foods may not be culturally tailored to communities they serve, and (c) processed foods, including sugar-sweetened beverages and snack bars, are among the "healthy" foods that owners and managers recognize in their stores.

Future interventions will need to address these conceptions of healthy foods to be effective.

Second, our study highlights the need to address the financial implications of public health interventions for food-stores' bottom line. The central theme in respondents' explanations of their business is that, as one store manager put it, "sales is what drives it". At first glance, the centrality of sales seems obvious. Bleich [59] argues, however, that public health researchers have not adequately addressed the priority of making sales. In particular, she suggests, existing food-store interventions (a) do not include sufficient fiscal measures to demonstrate short- and long-term impacts on profitability and (b) do not simultaneously consider the diverging goals of business owners and public health researchers [59]. Our results underscore the urgency of this critique. Store owners and managers in our sample articulated the primacy of making sales as the basis for all decision-making and acknowledged that what is good for business is not necessarily good for public health. Future interventions need to accommodate this perspective and provide better evidence regarding the fiscal impacts of store-based interventions. Without such evidence, it is unclear whether interventions are scalable, and it will be difficult to secure necessary buy-in from food-store owners.

Third, our findings extend multilevel, ecological models of the food environment. Story et al. [26] propose an ecological model of influences on what people eat, including individual-level factors, the social environment, the physical environment, and macro-level structures and policies. Our results suggest that the model applies not only to what people eat but also to what is available in food stores. Individual-level factors evident in store owners' and managers' interviews include perceived low consumer demand, perceived customer desire for convenience, parenting practices, and children's food preferences. Social environmental factors include household structure, neighborhood crime, and community mobilization for economic development. The physical environment includes store managers' and owners' perceptions of neighborhood conditions and the lack of space or infrastructure to stock healthy foods. Macro-level factors include the delocalization of decision making, the primacy of commercial interests over public health concerns, food prices, poverty, and overall economic conditions.

One implication of this framework is that efforts to reshape the consumer food environment require interventions across the spectrum of prevention. Escaron et al. [57] observe that the most successful store-based interventions combine multiple strategies (e.g., point-of-purchase information, advertising and promotion, pricing) to increase both the supply of and demand for healthful foods. Our study, based on food-store owners' and managers' understanding of their business, further suggests that store-based interventions would be enhanced by working systemically in other aspects of the food environment. For example, the delocalization of decision-making about what to sell in stores, which our study is the first to report, implies a need (a) to address corporate purchasing decisions and regional food-distribution chains and (b) to promote local sourcing of fresh foods [4]. Our respondents' perception that healthful foods cost more—in time as well as money—resonates with evidence regarding the association between parental employment and childhood obesity [60, 61], with the multiple meanings of price among consumers [62], and with agricultural and pricing policies that produce abundant low-cost, energy dense, and nutrient poor foods [63–65].

Finally, the ecological framework also encourages us to look beyond the food environment for potential solutions. Our respondents made clear that some of the biggest challenges they face have little to do with food. Rather, they identified "outside activities" such as theft, prostitution, and drug trafficking as factors that increase their costs and drive away potential customers. This finding corroborates a report from Nashville, where food store owners perceive crime as the most significant barrier they face [56]. Our respondents also linked their business practices to broader conditions such as the decline of neighborhood social integration and concentrated poverty. Their insight reminds us that, in addition to individual- and community-level health promotion, we must also confront social structures that produce unequal access to power and resources necessary to achieve health [66, 67].

We acknowledge several limitations of study design. The small sample size, although appropriate given our exploratory aims, limits our ability to make systematic comparisons among store types, and we are unable to assess whether store owners and managers who agreed to participate differ systematically from those who refused. We also lack data to test whether owners' and managers' perspectives are associated with the availability of healthful foods, sales, or customer characteristics. Last, our focus on food stores in low-income neighborhoods identified by Tallahassee's Community Redevelopment Agency (CRA) is justified by evidence of high need in the area [35, 36], but it limits our ability to generalize to other parts of the city or to explore how food-store owners' and managers' perspectives may vary by neighborhood-level socioeconomic status.

Despite these limitations, this study contributes new insight into the business practices and sociocultural processes that shape local food environments. Future efforts to explain and eliminate social inequalities in access to healthful foods will be strengthened by incorporating a better understanding of store owners' business priorities and perspectives, as well as the broader influences across multiple levels of analysis on the construction of consumer food environments.

ENDNOTE

[a]We include anonymous respondent identification codes (e.g., SS14) to make more transparent our use of the full data set in selecting illustrative excerpts.

ACKNOWLEDGEMENTS

This study was funded by a grant from the Blue Cross Blue Shield Foundation of Florida to the Leon County Health Department. The research is a project of the Health Equity Alliance of Tallahassee (HEAT), a community-academic partnership for action-oriented research on social inequalities in health (http://healthequityalliance.org/). The project would not have been possible without our team of research assistants: Krystina Gavin, Craig Lightfoot, Michael Robinson, Persell Scott, and Sejal Smith. We are especially grateful to the parents of the high school students on our team. Douglas A. Monroe assisted with interviewer training.

COMPETING INTERESTS

The authors declare that they have no competing interests.

AUTHORS' CONTRIBUTIONS

CCG contributed to the design of the study, developed data collection protocols, conducted data analysis, and drafted the manuscript. PQB and MMM helped to develop data collection protocols, participated in data collection, and supervised interviewers. AFS contributed to sampling design, data collection protocols, and preparing the manuscript. CB conceived of the study while working at the Leon County Health Department and contributed to the study design. All authors read and approved the final manuscript.

CONTRIBUTOR INFORMATION

Clarence C Gravlee, Email: ude.lfu@eelvargc.

P Qasimah Boston, Email: moc.loa@qakaraba.

M Miaisha Mitchell, Email: moc.loa@ahsiaimm.

Alan F Schultz, Email: ude.rolyab@ztluhcs_nala.

Connie Betterley, Email: moc.nsm@yelrettebjc.

REFERENCES

1. White House Task Force on Childhood Obesity . Solving the Problem of Childhood Obesity Within a Generation. Washington, DC: Executive Office of the President of the United States of America; 2010.
2. Moore L, Diez Roux A. Associations of neighborhood characteristics with the location and type of food stores. Am J Public Health. 2006;96:325–331. doi: 10.2105/AJPH.2004.058040.
3. Zenk S, Schulz A, Israel B, James S, Bao S, Wilson M. Neighborhood racial composition, neighborhood poverty, and the spatial accessibility of supermarkets in metropolitan Detroit. Am J Public Health.2005;95:660–667. doi: 10.2105/AJPH.2004.042150.
4. Larson N, Story M, Nelson M. Neighborhood environments: disparities in access to healthy foods in the U.S. Am J Prev Med. 2009;36:74–81. doi: 10.1016/j.amepre.2008.09.025.
5. Walker RE, Keane CR, Burke JG. Disparities and access to healthy food in the United States: a review of food deserts literature. Health Place. 2010;16:876–884. doi: 10.1016/j.healthplace.2010.04.013.
6. Lovasi GS, Hutson MA, Guerra M, Neckerman KM. Built environments and obesity in disadvantaged populations. Epidemiol Rev. 2009;31:7–20. doi: 10.1093/epirev/mxp005.
7. Laska M, Borradaile K, Tester J, Foster G, Gittelsohn J. Healthy food availability in small urban food stores: a comparison of four US

cities. Public Health Nutr. 2010;13:1031–1035. doi: 10.1017/S1368980009992771.

8. Black C, Moon G, Baird J. Health Place. 2014. Dietary inequalities: what is the evidence for the effect of the neighbourhood food environment?

9. Do DP, Dubowitz T, Bird CE, Lurie N, Escarce JJ, Finch BK. Neighborhood context and ethnicity differences in body mass index: a multilevel analysis using the NHANES III survey (1988–1994) Econ Hum Biol. 2007;5:179–203. doi: 10.1016/j.ehb.2007.03.006.

10. Holsten JE, Compher CW. Children's food store, restaurant, and home food environments and their relationship with body mass index: a pilot study. Ecol Food Nutr. 2012;51:58–78. doi: 10.1080/03670244.2012.635572.

11. Ford PB, Dzewaltowski DA. Disparities in obesity prevalence due to variation in the retail food environment: three testable hypotheses. Nutr Rev. 2008;66:216–228. doi: 10.1111/j.1753-4887.2008.00026.x.

12. Glanz K, Sallis JF, Saelens BE, Frank LD. Healthy nutrition environments: concepts and measures. Am J Health Promot. 2005;19:330–333. doi: 10.4278/0890-1171-19.5.330.

13. Caspi CE, Sorensen G, Subramanian SV, Kawachi I. The local food environment and diet: a systematic review. Health Place. 2012;18:1172–1187. doi: 10.1016/j.healthplace.2012.05.006.

14. Galvez MP, Pearl M, Yen IH. Childhood obesity and the built environment. Curr Opin Pediatr.2010;22:202–207. doi: 10.1097/MOP.0b013e328336eb6f.

15. Holsten JE. Obesity and the community food environment: a systematic review. Public Health Nutr.2009;12:397–405.

16. Sharkey J, Horel S. Neighborhood socioeconomic deprivation and minority composition are associated with better potential spatial access to the ground-truthed food environment in a large rural area. J Nutr.2008;138:620.

17. Izumi BT, Zenk SN, Schulz AJ, Mentz GB, Wilson C. Associations between neighborhood availability and individual consumption of dark-green and orange vegetables among ethnically diverse adults in Detroit. J Am Diet Assoc. 2011;111:274–279. doi: 10.1016/j.jada.2010.10.044.

18. Jennings A, Welch A, Jones AP, Harrison F, Bentham G, van Sluijs EMF, Griffin SJ, Cassidy A. Local food outlets, weight status, and dietary intake: associations in children aged 9–10 years. Am J Prev Med.2011;40:405–410. doi: 10.1016/j.amepre.2010.12.014.

19. Macdonald L, Ellaway A, Ball K, Macintyre S. Is proximity to a food retail store associated with diet and BMI in Glasgow, Scotland? BMC Public Health. 2011;11:464. doi: 10.1186/1471-2458-11-464.

20. Shier V, An R, Sturm R. Is there a robust relationship between neighbourhood food environment and childhood obesity in the USA? Public Health. 2012;126:723–730. doi: 10.1016/ j.puhe.2012.06.009.

21. An R, Sturm R. School and residential neighborhood food environment and diet among California youth.Am J Prev Med. 2012;42:129–135. doi: 10.1016/j.amepre.2011.10.012.

22. Lee H. The role of local food availability in explaining obesity risk among young school-aged children.Soc Sci Med. 2012;74:1193–1203. doi: 10.1016/j.socscimed.2011.12.036.

23. Morland KB, Evenson KR. Obesity prevalence and the local food environment. Health Place.2009;15:491–495. doi: 10.1016/j. healthplace.2008.09.004.

24. Leung CW, Gregorich SE, Laraia BA, Kushi LH, Yen IH. Measuring the neighborhood environment: associations with young girls' energy intake and expenditure in a cross-sectional study. Int J Behav Nutr Phys Act. 2010;7:52. doi: 10.1186/1479-5868-7-52.

25. Glanz K. Measuring food environments: a historical perspective. Am J Prev Med. 2009;36:S93–S98. doi: 10.1016/j.amepre.2009.01.010.

26. Story M, Kaphingst K, Robinson-O'Brien R, Karen G. Creating healthy food and eating environments: policy and environmental approaches. Annu Rev Public Health. 2008;29:253–272. doi: 10.1146/annurev.publhealth.29.020907.090926.

27. Cheadle A, Psaty B, Curry S, Wagner E, Diehr P, Koepsell T, Kristal A. Community-level comparisons between the grocery store environment and individual dietary practices. Prev Med. 1991; 20:250–261. doi: 10.1016/0091-7435(91)90024-X.

28. Zenk S, Schulz A, Hollis-Neely T, Campbell R, Holmes N, Watkins G, Nwankwo R, Odoms-Young A. Fruit and vegetable intake in African Americans: income and store characteristics. Am J Prev Med.2005;29:1–9. doi: 10.1016/j.amepre.2005.03.002.

29. Bodor J, Rose D, Farley T, Swalm C, Scott S. Neighbourhood fruit and vegetable availability and consumption: the role of small food stores in an urban environment. Public Health Nutr. 2008;11:413–420. doi: 10.1017/S1368980007000493.

30. Bodor JN, Ulmer VM, Dunaway LF, Farley Thomas A, Rose D. The rationale behind small food store interventions in low-income urban neighborhoods: insights from New Orleans. J Nutr. 2010;140:1185–1188. doi: 10.3945/jn.109.113266.

31. Gittelsohn J, Rowan M, Gadhoke P. Interventions in small food stores to change the food environment, improve diet, and reduce risk of chronic disease. Prev Chronic Dis. 2012;9:E59.

32. Dannefer R, Williams DA, Baronberg S, Silver L. Healthy Bodegas: increasing and promoting healthy foods at corner stores in New York City. Am J Public Health. 2012;102:e27–e31. doi: 10.2105/AJPH.2011.300615.

33. Rosecrans AM, Gittelsohn J, Ho LS, Harris SB, Naqshbandi M, Sharma S. Process evaluation of a multi-institutional community-based program for diabetes prevention among First Nations. Health Educ Res.2007;23:272–286. doi: 10.1093/her/cym031.

34. Gittelsohn J, Franceschini M, Rasooly I, Ries A, Ho L, Pavlovich W, Santos V, Jennings S, Frick K. Understanding the food environment

in a low-income urban setting: implications for food store interventions.J Hunger Environ Nutr. 2008;2:33–50. doi: 10.1080/19320240801891438.

35. Rigby S, Leone AF, Kim H, Betterley C, Johnson MA, Kurtz H, Lee JS. Food deserts in Leon County, FL: disparate distribution of Supplemental Nutrition Assistance Program–accepting stores by neighborhood characteristics. J Nutr Educ Behav. 2012;44(6):539–547. doi: 10.1016/j.jneb.2011.06.007.

36. Leone AF, Rigby S, Betterley C, Park S, Kurtz H, Johnson MA, Lee JS. Store type and demographic influence on the availability and price of healthful foods, Leon County, Florida, 2008. Prev Chronic Dis.2011;8:A140.

37. Israel B, Schulz A, Parker E, Becker A. Review of community-based research: assessing partnership approaches to improve public health. Annu Rev Public Health. 1998;19:173–202. doi: 10.1146/annurev.publhealth.19.1.173.

38. Barnes A, Roberts A. Tallahassee, Florida. Charleston, SC: Arcadia Publishing; 2000.

39. Rabby G. The Pain and the Promise: The Struggle for Civil Rights in Tallahassee, Florida. Athens, GA: University of Georgia Press; 1999.

40. Romney A, Weller S, Batchelder W. Culture as consensus: a theory of culture and informant accuracy.Am Anthropol. 1986;88:313–339. doi: 10.1525/aa.1986.88.2.02a00020.

41. Guest G, Bunce A, Johnson L. How many interviews are enough? An experiment with data saturation and variability. Field Methods. 2006;18:59–82. doi: 10.1177/1525822X05279903.

42. Bernard HR. Research Methods in Anthropology: Qualitative and Quantitative Approaches. 5. Lanham, MD: AltaMira Press; 2011.

43. Borgatti SP. Elicitation Techniques for Cultural Domain Analysis. In: Schensul JJ, LeCompte MD, editors. The Ethnographer's Toolkit. Volume Vol. 3. Walnut Creek, CA: AltaMira Press; 1998. pp. 1–26.

44. Gravlee CC, Bernard HR, Maxwell CR, Jacobsohn A. Mode effects in free-list elicitation: comparing oral, written, and web-based data collection. Soc Sci Comput Rev. 2013;31:119–132. doi: 10.1177/0894439312455312.

45. McLellan E, MacQueen K, Neidig J. Beyond the qualitative interview: data preparation and transcription. Field Methods. 2003;15:63–84. doi: 10.1177/1525822X02239573.

46. Borgatti S. ANTHROPAC 4.0 Methods Guide. Natick, MA: Analytic Technologies; 1996.

47. Wutich A, Gravlee CC. Water Decision-Makers in a Desert City: Text Analysis and Environmental Social Science. In: Vaccaro I, Smith EA, Aswani S, editors. Environmental Social Sciences: Methods and Research Design. New York: Cambridge University Press; 2010. pp. 188–211.

48. Ryan G, Weisner T. Content Analysis of Words in Brief Descriptions: How Fathers and Mothers Describe Their Children. In: de Munck VC, Sobo EJ, editors. Using Methods in the Field: A Practical Introduction and Casebook. Walnut Creek, CA: AltaMira Press; 1998. pp. 57–68.

49. Ryan G, Bernard H. Techniques to identify themes. Field Methods. 2003;15:85–109. doi: 10.1177/1525822X02239569.

50. MacQueen K, McLellan E, Kay K, Milstein B. Codebook development for team-based qualitative analysis. Field Methods. 1998;10:31–36. doi: 10.1177/1525822X980100020301.

51. Cohen L, Chavez V, Chehimi S. Prevention Is Primary: Strategies for Community Well Being. San Francisco: Jossey-Bass; 2007.

52. Ho LS, Gittelsohn J, Rimal R, Treuth MS, Sharma S, Rosecrans A, Harris SB. An integrated multi-institutional diabetes prevention program improves knowledge and healthy food acquisition in northwestern Ontario First Nations. Health Educ Behav. 2008;35:561–573. doi: 10.1177/1090198108315367.

53. Gittelsohn J, Suratkar S, Song HJ, Sacher S, Rajan R, Rasooly IR, Bednarek E, Sharma S, Anliker JA. Process evaluation of Baltimore Healthy Stores: a pilot health intervention program with supermarkets and corner stores in Baltimore City. Health Promot Pract. 2010;11:723–732. doi: 10.1177/1524839908329118.

54. Adams J, Halligan J, Watson DB, Ryan V, Penn L, Adamson AJ, White M. The Change4Life convenience store programme to increase retail access to fresh fruit and vegetables: a mixed methods process evaluation. PLoS One. 2012;7:e39431. doi: 10.1371/journal.pone.0039431.

55. Gardiner B, Blake M, Harris R, Gee C, Charaktis S, Choong C, Lade R, Duff L, Palermo C. Can small stores have a big impact? A qualitative evaluation of a store fruit and vegetable initiative. Health Promot J Aust. 2013;24:192–198. doi: 10.1071/HE13045.

56. Larson C, Haushalter A, Buck T, Campbell D, Henderson T, Schlundt D. Development of a community-sensitive strategy to increase availability of fresh fruits and vegetables in Nashville's urban food deserts, 2010–2012. Prev Chronic Dis. 2013;10:130008. doi: 10.5888/pcd10.130008.

57. Escaron AL, Meinen AM, Nitzke SA, Martinez-Donate AP. Supermarket and grocery store–based interventions to promote healthful food choices and eating practices: a systematic review. Prev Chronic Dis.2013;10:120156. doi: 10.5888/pcd10.120156.

58. Gittelsohn J, Kim EM, He S, Pardilla M. A food store-based environmental intervention is associated with reduced BMI and improved psychosocial factors and food-related behaviors on the Navajo Nation. J Nutr. 2013;143:1494–1500. doi: 10.3945/jn.112.165266.

59. Bleich SN. Generating better evidence to engage local food outlets in obesity prevention research. Prev Med. 2013;57:265–267. doi: 10.1016/j.ypmed.2013.07.015.

60. Cawley J, Liu F. Maternal employment and childhood obesity: a search for mechanisms in time use data.Econ Hum Biol. 2012;10:352–364. doi: 10.1016/j.ehb.2012.04.009.

61. Anderson PM. Parental employment, family routines and childhood obesity. Econ Hum Biol.2012;10:340–351. doi: 10.1016/j.ehb. 2012.04.006.

62. DiSantis KI, Grier SA, Odoms-Young A, Baskin ML, Carter-Edwards L, Young DR, Lassiter V, Kumanyika SK. What "Price" means when buying food: insights from a multisite qualitative study with Black Americans. Am J Public Health. 2013;103:516–522. doi: 10.2105/AJPH.2012.301149.

63. Cawley J. The economics of childhood obesity. Health Aff. 2010;29:364–371. doi: 10.1377/hlthaff.2009.0721.

64. Gereffi G, Lee J, Christian M. US-based food and agricultural value chains and their relevance to healthy diets. J Hunger Environ Nutr. 2009;4:357–374. doi: 10.1080/19320240903321276.

65. Nghiem N, Wilson N, Genç M, Blakely T. Understanding price elasticities to inform public health research and intervention studies: key issues. Am J Public Health. 2013;103:1954–1961. doi: 10.2105/AJPH.2013.301337.

66. Lane S, Keefe R, Rubinstein R, Levandowski B, Webster N, Cibula D, Boahene A, Dele-Michael O, Carter D, Jones T, Wojtowycz M, Brill J. Structural violence, urban retail food markets, and low birth weight.Health Place. 2008;14:415–423. doi: 10.1016/j.healthplace.2007.08.008.

67. Farmer P, Nizeye B, Stulac S, Keshavjee S. Structural violence and clinical medicine. PLoS Med.2006;3:e449. doi: 10.1371/journal. pmed.0030449.

Chapter 7

A review of literature on restaurant menus: Specifying the managerial issues

Bahattin Ozdemir, Osman Caliskan'

Gastronomy and Culinary Arts, Akdeniz University Tourism Faculty, Dumlupinar Bulvarı, Konyaaltı, Antalya07058, Turkey

ABSTRACT

This conceptual study specifically aims at reviewing the critical managerial issues of menu, and demonstrating the conceptual structure of menu management. Based on the conceptual and empirical findings of menu literature, the major menu management issues are menu planning, menu pricing, menu designing, menu operating and menu development. Additionally, the paper makes a discussion on the conceptual relations between menu and meal experience. Given the scarcity of research that incorporates evidences and concepts of previous studies in one single study, the conceptual structure of menu management presented in this paper allows a comprehensive understanding of menu and forms a theoretical basis for future research.

KEYWORDS

Menu; Menu management; Restaurant firms

1. INTRODUCTION

Menu maintains its dominant position in restaurant firms, since it is the core of food and beverage operations. Specifically, menu is an instructor that clearly dictates (i) what will be produced, (ii) what type of equipment and ingredients are needed, and (iii) which qualifications employees should have. Menu also functions as a communicating and selling tool (Kincaid and Corsun, 2003). More specifically, it communicates not only the food and beverage offerings, but also the image of the firm. Furthermore, menu is a base on which the customers make their food choices and a well designed menu can direct customers' attention to the items the firm wants to sell more (Antun and Gustafson, 2005).

The critical qualifications attributed to menu describe its importance to restaurant operations and customer experiences. In fact, in contemporary understandings of menu, it has material and immaterial meanings. Materially, menu is a list or a card which documents the food and beverage options being offered by a restaurant. Immaterial meanings of menu highlight that it is a medium that affects customers' perceptions of restaurant experience (Wansink et al., 2005), and it is also an element of restaurant management that requires conducting critical managerial activities such as planning, pricing, designing and analyzing (Jones and Mifli, 2001). It also differs from the main product of a restaurant. Prior research (Johns and Kivela, 2001, Gustafsson, 2004 and Hansen et al., 2005) claim that the primary product of restaurants is meal experience and it is constituted by many components such as food and beverages, atmosphere, social factors and management. What customers receive from consumption in restaurants is their memories and overall perceptions of meal experience, and menu is part of that experience and has associations with customers' perceptions. Consequently, menu cannot be considered as the overall product of a restaurant but it is an important component complementing atmosphere, interactions with servers and other customers, and managerial systems which are jointly available in a restaurant where a good, memorable meal experience is the main product.

Since it is a key factor in attracting and satisfying customers and effecting performance of restaurant firms, menu has been gaining a considerable academic attention from researchers. The previous studies on menu constitute a substantial body of literature and they primarily examined the issues including menu planning, menu pricing, menu designing, and menu analysis. For instance, Kivela (2003) investigated menu planning through employing an experimental research design and proposed a menu item selection framework. Menu pricing was examined by several authors from a range of perspectives including the price perceptions of customers (Shoemaker, 1994 and Iglesias and Guillen, 2002), price sensitivity, pricing methods (Kelly et al., 1994,Raab et al., 2009 and Kelly et al., 2009), and psychological pricing (Carmin and Norkus, 1990, Naipaul and Parsa, 2001 and Parsa and Hu, 2004). Menu designing has also gained a particular attention from menu researchers who have particularly focused on positioning the menu item on the menu card (Bowen and Morris, 1995, Kincaid and Corsun, 2003 and Reynolds et al., 2005), labeling (Lockyer, 2006 and Wansink et al., 2001) and describing menu items (McCall and Lynn, 2008 and Hwang and Lorenzon, 2008). In the field of menu analysis, researchers have offered several models such as menu engineering (Kasavana and Smith, 1982), analysis of menu items' profitability (Hayes and Huffman, 1985), and more recently multi-dimensional analysis of menu items (Taylor and Brown, 2007). From a practical perspective, menu is directly related to many functions of food and beverage operations and also absorbs a wide range of interrelated managerial decisions in the areas of planning, pricing, designing, operating, and developing menus.

Despite the growing academic attention to restaurant menus, previous research has a certain drawback. Principally, there is an apparent lack of an attempt at incorporating the major thoughts and ideas of menu research into one single study. Given the above background, this conceptual study presents a review of relevant literature to frame the current discussion on menu with an attempt to identify critical managerial issues. Thus, the specific objectives of the current study are (i) to identify the critical managerial issues of menu, and (ii) to demonstrate the conceptual structure of menu management. As the menu is a key to influencing success of firms in foodservice industries, this study could assist restaurateurs in understanding the role of menu in their managerial decisions. From an academic perspective, this paper presents a holistic framework for menu researchers since it incorporates the conceptual and empirical findings of previous menu research into one single study.

2. LITERATURE REVIEW

2.1 Theoretical underpinnings

The perspectives of organizational, marketing and strategic management theories provide a reliable theoretical groundwork to understand the important managerial aspects of menu. For instance, organization theory explicitly emphasizes the influence of external environment both on the decisions of firm managers and the survival of firms on the long run. More specifically, external environment is one of the central themes of organization theory and the relevant studies (Duncan, 1972, Jurkovich, 1974, Dess and Beard, 1984 and Ashill and Jobber, 1999) primarily propose that managers essentially consider the influential external factors that create uncertainty, diversity and volatility while making their decisions. Planning and operating menus in a restaurant context involve considering external factors such as customers, rivals, and vendors that have a great potential in creating uncertainty, diversity and volatility in the restaurants' immediate business environment. Complementing this view, marketing theory recognizes the importance of identifying the needs and expectations of customers, and developing and improving products and service perfectly fit to those needs and expectations. Additionally, pricing, promoting and distributing the products and services should also be consistent with the customers' needs and expectations on the one hand, and with the firm's objectives on the other hand. Thus, in the restaurant context, it is imperative that menu as the food and beverage combinations offered by a restaurant reflects the expectations and needs of customers. Moreover, managing menus involves planning, pricing, designing, distributing and promotional decisions which are also the main issues of marketing.

Strategic management is another promising theoretical perspective that helps us understand the importance of menu and its associated managerial activities with reference to competitiveness of restaurants. Indeed, creating and sustaining a competitive advantage is the main focus of strategic management discipline. More specifically, strategic management principally investigates the phenomenon of sustainable competitive advantage as a source of performance differences among business firms (Nerur et al., 2008). Following this argument, it is plausible to consider the role of menu in creating sustainable competitive advantage for restaurants. In particular, offering a unique meal experience and assuring customer satisfaction are the main sources of competitive advantage for a restaurant. As one of the important components of meal experience, menu deserves a special attention to understand how a restaurant firm can achieve its strategic goals, can create a sustainable competitive advantage, can satisfy its customers and can outperform the rivals.

Strategic management literature currently encompasses a wide range of strategic management schools posing different strategy definitions and processes (Sarvan et al., 2003). As a result, there are various understandings of the meaning of strategy (Okumus, 2002). For example, Mintzberg et al. (1998) define 10 schools of strategic thinking as design, planning, positioning, entrepreneurial, cognitive, learning, power, cultural, environmental and configuration schools. Among the 10 strategy schools, positioning school provides a clear typology of strategy alternatives. For instance, Porter (1980)proposes well-known three generic strategies: low cost leadership, differentiation and focus strategies, those are also available for restaurant firms. When considered in a restaurant context, each strategic alternative has associated implications for menu. For example, if the strategy is low cost leadership, then decisions on planning, pricing and designing menu reflect the strategy through selecting low cost menu items, pricing menu items at a lower price than rivals, and attractively presenting menu items with low prices on menu card, which probably result in customers' perceptions of a satisfactory meal experience with a reasonable quality. Conversely, when a differentiation strategy is chosen, a restaurant manager is expected to be attentive to (i) selecting or innovating unique menu items, (ii) setting higher prices with an expectation that customers are willing to pay premium for a differentiated product, and (iii) heavily concentrating on attractively presenting unique items on menu card to create a positive image of a unique meal experience. Thus, all managerial decisions with reference to planning, pricing and designing menus are the reflections of the chosen strategy such as low cost leadership or differentiation.

2.2 Menu management issues

Menu management is a relatively new concept which is not widely recognized by researchers or practitioners. Indeed, in the relevant literature very few studies have used the concept without providing a certain definition of it. Thus, the major aim of literature review in this section is to trace the variables investigated by menu researchers. The review purposely restricted to the journal articles which were subjected to a rigorous academic judgment process before publishing.

The previous research includes empirical and conceptual studies examining various managerial aspects of menu. Table 1 demonstrates the examples of those studies and their variables. As seen in Table 1, there are different sets of variables corresponding to one or several of the menu management issues including menu planning, menu pricing, menu designing, and menu development. Indeed, among the articles referred in Table 1, the empirical study of Jones and Mifli (2001), which also forms the conceptual base for the current study, jointly investigates planning, designing, developing, pricing and analyzing issues in seven hospitality firms. Jones and Mifli (2001: 61–62) relate menu planning to identifying the product range of restaurant firm, and menu design to the creation of menu card. They define menu development as subsequent alteration of the planned menu. Finally, menu analysis is associated with a range of techniques and procedures employed in order to analyze the menu performance. The following section elaborates on menu management issues by drawing evidence from the menu literature.

2.2.1 Menu planning

Recent studies on menu planning examine variables such as menu item selection process and criteria, menu changes and variety, and menu item innovation. Those studies primarily assume that menu planning is a critical managerial activity to the success of restaurant firms. In this regard, Kivela (2003) describes menu planning as a menu item selection process. After heavily criticizing the traditional approach to menu planning which mainly relies on executive chefs' gastronomic knowledge, he proposes a Menu Planning Qualitative Variables (MPQV) approach which considers gastronomic, financial, marketing and menu mechanics variables. Morrison (1997) and Kwong (2005)also share the idea that menu planning is a process of selecting menu items and their research reports empirical evidences that the criteria of menu item selection are based on the cost of raw materials, cost of labor, profitability of menu item, the skill level of staff needed, availability of ingredients, space and equipment requirements, and customer demand. A valid alternative to selecting menu items from an existing items' pool based on some pre-determined criteria is to make menu item innovations. Recently, some researchers (Ottenbacher and Harrington, 2007 and Ottenbacher and Harrington, 2009) have focused on innovation processes in restaurant kitchens. The studies of those researchers empirically demonstrate how chefs create new menu items relying on innovation processes that are generally encompassing the stages such as idea generation, screening, concept development and tests, commercialization, implementation and evaluation.

Glanz et al. (2007) emphasize that changing menu's content is also an important component of menu planning and their study reveals that the major criteria of menu changes are whether the new menu item will attract new customers or maintain the existing customer base, and how it will affect sales and profit. Furthermore, menu changes may lead to menu variety, and according to findings presented by Bernstein et al. (2008), consumers value variety in their menu choices. After analyzing the recent trends in food preferences of consumers, DiPietro et al. (2004) conclude that healthy food items are becoming an increasingly important part of restaurant menus even in quick service segment. With respect to the growing trend towards healthy food choices, menu planning becomes more complicated due to associated activities for searching, selecting or innovating the healthy menu options.

Table 1. Examples of articles investigating menu issues.

Article	Main focus	Corresponding menu issue
Pavesic (1983)	Menu engineering model including weighted contribution margin and food cost percentage	Menu analyzing
Hayes and Huffman (1985)	Menu item profitability analysis and goal value analysis	Menu analyzing
Bayou and Bennett (1992)	Menu item profitability analysis based on analysis of market segments (four levels of profitability analysis)	Menu analyzing
Kelly et al. (1994)	The relationship between price elasticity and pricing	Menu pricing
Shoemaker (1994)	The effect of a decoy item on the price value perception of a menu	Menu pricing
Bowen and Morris (1995)	The relationship between menu item placement on the menu card and sales of menu items	Menu designing
LeBruto et al. (1995)	Menu engineering model including labor costs as a menu analysis method	Menu analyzing
Morrison (1997)	Development of menu engineering models and a study on upscale restaurants menu planning practices	Menu analyzing Menu planning
Jones and Mifli (2001)	Menu planning, pricing, designing, analyzing and improving practices of seven hospitality firms	Menu planning Menu pricing Menu designing Menu analyzing and improving
Naipaul and Parsa (2001)	The relationships between psychological pricing comprising odd and even number price endings, and value and quality perceptions of customers	Menu pricing
Wansink et al. (2001)	The relationship between descriptive menu item labels and customer perceptions	Menu designing
Iglesias and Guillen (2002)	Price perceptions of customers comprising monetary and non-monetary sacrifices	Menu pricing
Kincaid and Corsun (2003)	The relationship between the layout of menu item on menu card and item sales	Menu designing
Kivela (2003)	Menu planning approaches and menu item selection variables	Menu planning
DiPietro et al. (2004)	Consumer attitudes towards healthy menu items and investigation of sales trends in healthy menu items	Menu planning
Antun and Gustafson (2005)	Menu design characteristics and pricing strategies used by restaurants and clubs	Menu pricing Menu designing
Kwong (2005)	Asian restaurants' practices of menu engineering and menu design	Menu analyzing Menu designing

Cohen et al. (2006)	The menu item dependency and the effect of loss-leader pricing on menu items performances, and proposal of a multidimensional menu analysis technique	Menu analyzing
Lockyer (2006)	The relationships between the wording of menu items and customers' menu item choice	Menu designing
Taylor and Brown (2007)	A review of menu analysis literature including improvements to menu item performance	Menu analyzing and improvements
Ottenbacher and Harrington (2007)	The menu item innovation process in fine dining restaurants	Menu planning
Bernstein et al. (2008)	The relationship between menu variability and consumers' restaurant preferences	Menu planning
Hwang and Lorenzon (2008)	Relationships among nutrition labeling of menu items, consumer perceptions and menu pricing	Menu pricing Menu designing
McCall and Lynn (2008)	The relationship between the complexity of menu item descriptions and potential customers' evaluations	Menu designing
Raab et al. (2009)	Pricing approaches adopted by hospitality firms and price sensitivity measurement as a pricing method	Menu pricing
Taylor et al. (2009)	Proposal of data envelopment analysis as menu analysis technique	Menu analyzing
Ottenbacher and Harrington (2007)	The menu item innovation process in quick service restaurants	Menu planning

Menu planning researchers have contributed to establishing the conceptual structure of menu planning by introducing and exploring menu item selection, menu item innovation, menu variety, menu change and menu content variables. It seems that research on menu planning largely holds management theories' perspective and highlights the effects of external (customer, rivals, vendors) and internal (cost structure, employees, equipment) factors on menu planning decisions of managers. Considering the findings and suggestions of the previous studies, it is possible to make an operational definition of menu planning. In this regard, menu planning refers to menu item selection and/or menu item innovation with a consideration of menu variety and menu change in order to generate a menu content of an optimum composition of food and beverage offerings that are most likely to gratify both the customers' and restaurant firm's expectations.

2.2.2 Menu pricing

Menu pricing is a quite popular topic among menu researchers and it seems that a wide range of variables such as price perceptions of customers, price sensitivity, price elasticity, pricing methods and specifically the psychological pricing have attracted their attention. For instance, Iglesias and Guillen (2002) specifically underline the importance of customers' perceptions of menu item prices and suggest that total price perception of customers consists of perceived monetary price and perceived non-monetary price. Additionally, they indicate that customers also consider reference price and compare it to the actual price of menu item when making purchase decisions. Specifically, to some extent the assessment of the price of a menu item is based on the prices of alternatives. In a restaurant setting, for example, the prices of decoy items on the menu card may act as reference prices (Shoemaker, 1994). Depending on the price sensitivity or price elasticity, customer demand to food and beverage offerings of a particular restaurant may change in response to an increase or decrease in the prices (Kelly et al., 1994 and Raab et al., 2009). Moreover, presentation of price endings can affect customers' value or quality perceptions of associated menu items when they are exposed to psychological pricing methods (Naipaul and Parsa, 2001).

The researchers in the field of menu pricing install a marketing theory perspective into menu management through emphasizing the perceptions and expectations of customers. Following the conceptual background provided by previous research, menu pricing can be described as the formulation of menu item prices in response to both sales, cost and profit objectives of the firm, and perceptions and expectations of customers. Additionally, as suggested by the relevant literature, setting the menu item prices is a complicated task since it requires becoming attentive to complex effects of the interrelated factors such as customers' perceptions, price elasticity and psychological pricing.

2.2.3 Menu designing

Researchers suggest that menu designing comprises managerial decisions regarding design of menu card characteristics, position, label, and description of menu items on the menu display (card or board). More specifically, the previous research on menu design has associated the menu item's position, label and description with item sales and customers' perceptions.

Although there are theoretical explanations asserting that the menu card contains some sweet spots and a particular menu item placed on a sweet spot generates higher sales; the empirical evidences reported by Bowen and Morris (1995), Kincaid and Corsun (2003), Reynolds et al. (2005) do not support that assumption. However, recently Choi et al. (2010) and Yang (2012) report evidences that restaurant menus may consist of sweet spots where the first gazes of customers focus on. Some other researchers present empirical findings that verify the association between an item's sales and its label and description. For instance, the findings of Wansink et al. (2001) and Lockyer (2006)evidence that labeling or wording of menu items' names significantly influences the customers' menu item choice and perceptions. More specifically, according to Wansink et al. (2005), evocative labels on a menu card attract customers' attention to specific items and positively affect their taste and value perceptions of those items. Additionally, the empirical study of McCall and Lynn (2008) demonstrate the importance of item descriptions on a menu card through the finding that the complexity of menu item descriptions positively affects the customers' evaluations. Shoemaker et al. (2005)present a similar finding and conclude that details provided in items' descriptions have a positive impact on customers' value perceptions. Alternatively, Hwang and Lorenzon (2008) advocate that nutritional information provided on the menu card can enhance the positive attitudes of customers by informing them about healthy options in the menu and further make them change their orders to healthy items. Menu design also relates to a variety of factors including background and text colors, texture, photos, fonts, the use of boxes, and menu size as characteristics of menu card (Reynolds et al., 2005) which also have considerable effects on customers' item choice as empirically supported byGuéguen et al. (2012) who found that pictures of sea on the menu card significantly increase the consumption of fish dishes.

The studies focusing on menu design seem to make considerable contribution to our understanding of menu management through clarifying the relationships between item sales and menu design variables. In this way, they introduce a merchandising approach to menu management and hold a marketing perspective which assists managers in attracting customer attention to particular menu items. Given the above background, menu design refers to designing menu card with respect to menu card characteristics, menu item layout, menu item descriptions and labels in order to create an attractive menu that not only provides information but also directs customers' attention to the items that the firm wants to sell more.

2.2.4 Menu operating

Menu operating is essentially included in menu management issues despite the fact that it very rarely appears in Table 1. In fact, it is not possible to state that menu researchers have paid particular attention to operations of menu except Jones and Mifli (2001). However, the studies on subjects such as sales forecasting, budgeting, cost control, service failures, and food safety provides insightful contribution to our understanding of menu operating.

In particular, menu operating involves management of menu items' production and service processes. Thus, it is expected that restaurant managers frequently consider three key factors as quality, hygiene and cost while making decisions and taking actions of menu operating. Accordingly, standardizing production, minimizing service failures, assuring food safety, forecasting demand to menu items, making budgets and controlling costs become critical activities of operational management of menus. These are also the major considerations of quality management approaches such as ISO: 9001 and ISO 22000 which are also adaptable to restaurant operations. When considered in the restaurant context, one of the main components of such a quality management system will be assurance of food quality and safety relying on standard production, minimum service failures, optimum demand and suitable cost structures; those are briefly explained below with reference to relevant literature.

In the production process of menu items, development and implementation of some standards such as standard yields, standard purchasing specifications, standard recipes and standard portions are useful tools for sustaining a quality level (Davis and Stone, 1998). As found by Jones and Mifli (2001), especially in the chain restaurants portion control, standard recipes, quality control, and standard specifications are integral components of menu operations.

Serving a meal in a restaurant (especially in a table service restaurant) is composed of several sequential service stages such as greeting and seating, order taking and delivery, consumption, and payment and exiting (Namkung et al., 2011). Inevitably, service failures may occur at any stage of service, and these failures may lead to negative customer perceptions. For example, customers may complain following a failure (Namkung et al., 2011). Service failures may also negatively affect the customers' satisfaction levels and their likelihood of returning (Namkung and Jang, 2010). Consequently, menu operating decisions of managers need to consider serving menu items properly without any delays or defects in order to minimize service failures at all service stages.

Restaurant customers are becoming increasingly concerned about possible risks stemming from unsafe food practices (Fatimah et al., 2011) and food safety is a critical restaurant selection attribute (Worsfold, 2006). Thus, in menu operating stage restaurant managers are challenged by an urgency to produce and serve safe food. In order to ensure food safety, restaurant managers need to recognize the potential biological, physical and chemical hazards and take preventive actions to minimize the occurrence of these hazards (Fatimah et al., 2011). However, restaurant industry frequently faces food safety problems that mainly occur due to unsafe food sources, inadequate cooking, improper food handling, improper storage temperatures, contaminated equipment and poor personal hygiene (Sanlier et al., 2010). As a solution to the current food safety problems, it is widely advised that restaurant firms should implement a food safety management system such as HACCP (Hazard Analysis and Critical Control Points) which consists of principles and practices that encompass all aspects of processing, preparing, handling and serving food in a completely safe way (Eves and Dervisi, 2005).

Sales forecasts have a key role in operational management of menu due to the perishable nature of food products. More specifically, an accurate prediction of demand to menu items ensures the smooth functioning of production and service without falling behind the demand, which may cause customer dissatisfaction, or exceeding the demand, which increases the costs (Ryu et al., 2003). Previous studies on sales forecasting have documented the available forecasting methods such as simple moving average method, simple exponential smoothing method, time-series models, regression, and Delphi etc. (Green and Weaver, 2008). Some researchers suggest the use of those quantitative techniques rather than merely intuitive assessments of managers since naïve intuitive approaches produce less accurate forecasts than do quantitative techniques (Miller et al., 1991). Besides, Ryu et al. (2003) also highlight the importance of using simple and timesaving forecasting methods.

In menu operating, budgets and cost control systems assist restaurant managers in assuring that there is no serious deviation from profit and cost objectives of restaurant firm in the entire process of procurement, storage, production and service. Underlying this notion, research by Jones (2008) evidences that hospitality firms produce budgets for cost control and performance evaluation purposes. Loss of profit due to the poor controls is a common concern in hospitality settings (Borchgrevink and Ancill, 2003). As a result, restaurant managers consider and implement a cost control system which encompasses purchasing, storage, issuing, production and service stages. The major aim of such a system is to minimize potential loss because of shrinkage, spoilage, spillage, waste and theft. However, as stated by Vaughn et al. (2010), the cost control system of many restaurant firms is based solely on the control of food costs. As a result, the restaurant has no knowledge of ultimate cost of menu items. Alternatively, Vaughn et al. (2010) suggest the use of activity based costing method in order to accurately determine and control the cost of menu items.

Given the above explanations, menu operating refers to smooth operation of menu with assistance of sales forecasts, standards, food safety practices, budgets and cost control systems in order to minimize deviations from quality, hygiene and cost objectives of restaurant firms.

2.2.5 Menu analysis

Researchers have particularly dealt with developing some quantitative models to analyze the performance of menu items. Additionally, some of them also proposed improvement strategies to eliminate the performance deficiencies of menu items. In the 1980s, early systematic menu analysis approaches based on matrixes were introduced by Miller (1980), Kasavana and Smith (1982) and Pavesic (1983). These matrix based models were criticized by subsequent studies and several weaknesses were noted (Taylor and Brown, 2007). Among them is the notion that indicator of menu items' profitability is only the contribution margin which is computed through subtracting food cost from selling price. Consequently, this approach does not include other variable and fix costs such as labor and energy. Paying special attention to this deficiency of matrix models, LeBruto et al. (1995) attempt to include labor costs into the menu engineering model. Moreover, Hayes and Huffman (1985) and Bayou and Bennett (1992), known as the developers of menu item profitability analysis, seek to identify profit of menu item by subtracting total cost from sales of the item. Recently, some multi-dimensional analysis approaches were developed (Taylor and Brown, 2007). For instance, Cohen et al. (2006)propose a multi-dimensional menu analysis model employing five performance dimensions including food cost, popularity, sales price, contribution margin, and weighted contribution margin. More recently, Taylor et al. (2009) suggest that data envelopment analysis is a useful multidimensional technique to analyze menu item performance.

Taylor and Brown (2007), Jones and Mifli (2001) and Kwong (2005) provide information about possible improvement strategies for menu item performance enhancement. Indeed, subsequent to the analysis of menu is the implementation of suitable improvement strategies for performance deficiencies. According to the authors, promotion and reposition of menu item; modification of presentation, recipe, and price are strategies for increases in the sales volume. Recosting, repricing, and recipe modification are optional solutions to cost problems. High performer menu items may be left as they are and low performers may be eliminated.

Studies on menu analysis reveal that evaluation of menu performance is mainly based on the financial performance indicators such as cost, popularity, profit and price. Additionally, improvements are made in order to increase sales or decrease costs. Therefore, it is obvious that menu analysis is made with respect to restaurant firms' financial goals. Relying on the arguments in the relevant literature, menu development may be proposed as a two dimensional concept including menu analysis and improvements instead of menu analysis concept alone. In this sense, menu development refers to analyzing menu item performance and then taking corrective actions to improve the performance imperfections of menu items.

3. DISCUSSION

It becomes clear that prior menu research has offered major menu management issues such as menu planning, menu pricing, menu designing, menu operating, and menu development. Critically, these menu management issues are not independent but interrelated. The success of any decision made in relation to a particular menu management issue largely depends on the performance rendered in relation to other issues. Furthermore, menu management is not a static managerial activity, but it is an ongoing process. This process has at least two main objectives including (i) satisfying the customers, and (ii) achieving the goals of restaurant firm. As stated by Kincaid and Corsun (2003), the menu contributes to restaurant firm's performance through generating profit for the firm and providing value and satisfaction to the customer. In this regard, menu management is a comprehensive concept that refers to an ongoing process comprising planning, pricing, designing, operating and developing the menu of a restaurant firm. Table 2 summarizes the conceptual structure of menu management along with the relevant variables extracted from literature.

Table 2. Conceptual structure of menu management.

Management issue	Short description	Outcome	Relevant variables
Menu planning	Process of generating an optimal menu content	Menu content	Menu item selection criteria Menu changes and variety Menu item innovation
Menu pricing	Formulation of menu item prices	Menu item prices	Pricing methods Price elasticity or sensitivity Price perceptions of customers Psychological pricing
Menu designing	Designing the menu display (menu card or board)	Menu display (menu card or board)	Menu card characteristics Menu item layout Menu item description Menu item label
Menu operating	Developing and implementing standards and procedures regarding quality, hygiene and cost	Operations	Budgeting Sales forecasting Assuring food safety Controlling costs Standard recipes Standard costs Standard yields Standard portions
Menu development	Analyzing performance of menu items and improving the performance imperfections	Menu items' performances and improvements	Analysis methods Improvement strategies

As seen in Table 2, planning, pricing, designing, operating, analyzing and improving menu include managerial decisions. Those decisions may have profound effects on customers' positive perceptions of the meal experience and their satisfaction with restaurant which further lead to a better performance of the restaurant firm. As meal experience is the primary product of restaurant firms (Johns and Kivela, 2001), then an assessment of associations between menu management issues and components of meal experience may contribute to our understanding of how menu might enhance the meal experience perceptions of customers.

A synthesis of existing literature (Gustafsson, 2004 and Hansen et al., 2005) reveals that components of meal experience are restaurant interior and atmosphere, social meeting, food and beverages, company of other people, and management control system. Since food is one of the most influential restaurant selection attributes (Kivela, 1997) and a determinant of customer satisfaction and retention (Kivela et al., 2000), food and beverages constitute the core of meal experience (Gustafsson, 2004). As menu management determines what will be the combination of food and beverages and their prices, how they will be presented on the menu card, and how they will be produced and served, it is possible to assume that menu has primary implications for the core component of meal experience. For instance, a menu planning activity attentive to customer expectations assures that customers can receive what they expect in terms of menu content, menu variety and changes. Similarly, an efficient use of menu pricing approaches can affect the way customers perceive the meal experience. Certain practices of presenting price-endings and considering customers' price perceptions while establishing prices play a key role on how customers evaluate the menu items and restaurant itself. For instance, in quick service settings price endings with “9” may generate a sense of value while “0” in price-endings may create a sense of quality in full-service settings (Naipaul and Parsa, 2001). Moreover, differentiating the menu items and accordingly their prices from potential competitors' offerings through innovations in dishes (Ottenbacher and Harrington, 2009) may assist in forming the customers' perception of a unique meal experience without any reference to prices of rivals' products (Iglesias and Guillen, 2002).

Existing menu design research suggests that customers' purchasing behavior and value and quality assessments can also be influenced by menu item labeling and description (Yang et al., 2009). For instance, menu item labels which awake positive emotions of customers, and descriptions including nutritional information may enhance the meal experience perceptions of customers. Specifically, labels and descriptions of menu items may lead customer to have positive perceptions of taste, value, health and quality aspects of food (Shoemaker et al., 2005; Wansink et al., 2001 and Wansink et al., 2005). Additionally, a restaurant manager who wishes to ensure that customers have a unique meal experience may consider that the unique food and beverage offerings of the establishment are located on the menu card where they are most probably seen and frequently ordered (Choi et al., 2010). Moreover, physical characteristics of menu card such as background, colors, texture, photos, fonts, size etc. may be designed in combination with the other atmospherics of the restaurant (décor, lighting, layout, table settings, appearance of service staff etc.) in order to generate an overall appealing image as the restaurant atmospherics have a considerable effect on customers' perception of meal experience, and their satisfaction and retention (Ryu and Han, 2011).

Operational decisions and actions related to menu such as standardizing and offering a consistent quality, minimizing service failures, accurately predicting demand to menu items, and assuring food safety are essential components of providing a prosperous meal experience. For instance, customers may be dissatisfied when their orders are delayed or cannot be served due to shortage of ingredients because of poor sales forecasting. Moreover, customers who are attentive to food safety issues may assess all visible cues such as cleanliness of utensils, serving temperature and personal hygiene of service staff in order to be certain the safety level of food and these assessments may influence their satisfaction with meal experience (Fatimah et al., 2011).

The management control system component of meal experience also deserves attention from menu management perspective. Sales, cost and profit are subjected to consideration in menu management. For instance, selecting menu items with reference to sales; cost and profit objectives of firm; increasing item sales with pricing and designing tactics; controlling the costs with assistance of sales forecasting, and budgeting and cost control systems are all related to management control system component of meal experience. Menu analysis also assists in management control system. As a periodical performance monitoring mechanism, menu analysis is undertaken in order to evaluate whether the menu management process is successful. Menu analysis requires recording and collecting menu items' performance data which are specific to the establishment and those data can be integrated into the restaurant's overall performance assessments. Menu analysis also contributes to enhancing the meal experience when it includes improvements in terms of making adjustments in menu planning, pricing, designing and operating in response to customers' expectations.

4. CONCLUSION

Relying on the previous studies on menu, this study shows that the menu literature is constituted by five menu management issues comprising menu planning, menu pricing, menu designing, menu operating and menu development. Menu management concept, then, is defined as an ongoing process including menu management issues.

4.1 Managerial implications

A review of menu research shows us that restaurant managers have to play diverse and inter-related roles in relation to menu management. In this sense, a restaurant manager should become a planner, a price setter, a designer, an operator and an analyst. On the one hand, these roles are diverse because each of them requires different tasks listed inTable 2 in this study. On the other hand, they are interrelated since each task should be accomplished in close relation to others. Thus, the challenging aspect of menu management is that managers essentially integrate the diversified and complex menu management issues and understand the interdependencies among them instead of treating each issue individually. In this way, managers may have an ability to use menu in enhancing the meal experience perceptions of customers and attaining the cost and profit objectives of restaurant firms.

As discussed in the current paper, menu may have a strong influence on customers' meal experience perceptions. The restaurants which aim to have a higher level of performance than their rivals essentially concentrate on creating a unique meal experience. Meal experience is a multi-component construct and restaurant managers need to find unique ways of combining the meal experience components and menu variables in order to provide a unique experience for their customers. Relying on the literature review presented above, it seems that one possible way to accomplish this is to treat menu management as a process consists of interconnected stages and following paragraphs include some major suggestions to managers explaining how to operate such a process.

Since, menu is a determinant of purchasing, receiving, storage, production and service requirements, a starting point for managers is to identify the current state of internal and external factors related to menu management. For instance, the equipment and staff portfolio of the firm in terms of qualitative and quantitative features should be overviewed because of the strong influence of menu on the equipment and staff requirements. The processes, cost structures, profit, sales and prices of rivals should be examined in order to assess and differentiate the firm's own offerings. Importantly, the needs and expectations of the existing and potential customers need to be captured and analyzed. Other environmental factors such as economical, social, political and legal issues and the possible changes in them should also be analyzed. In this regard, at the initial stage of menu management process there should be a situational analysis that comprises collecting and analyzing data related to external and internal variables such as customers, rivals, environmental factors, and the firm's capabilities. Consequently, the situational analysis stage forms a base which contains useful data to effectively perform the subsequent menu planning, pricing and designing stages of menu management process.

As stated by Reich (2002), a restaurant firm may have long term and short term objectives. The long term objectives comprise social responsibility, profitability, productivity, employee development, employee relations, competitive position and technology development. Short term objectives include increasing profitability, decreasing costs, increasing customer satisfaction, and increasing market share. Accordingly, it is reasonable to state that the menu management process should produce outcomes which are consistent with the long term and short term objectives of the restaurant firm. A review of the firm's long term and short term objectives provide insights to understand the expected outcomes of the menu management process. Thus, in the second stage following the situational analysis, managers should clearly identify the specific objectives of menu management process. The objectives specific to menu management are most likely to be the short term objectives such as profitability, cost levels, and customer satisfaction.

Menu planning, pricing and designing are interrelated issues. In order to accomplish the task of pricing, managers need to know the menu items, their contents and accordingly their costs. Similarly, menu designing requires knowing the names, content and prices of menu items. Furthermore, managers should base their menu planning, pricing, and designing decisions on the outcomes of situational analysis and review of objectives.

Subsequent to planning, pricing and designing, the planned menu is put into production. Therefore, operating the planned menu requires adopting standard procedures in order to achieve the expected performance determined in the planning stage. The menu operation should also consist of sales forecasting, food safety, budgeting, and cost control. Then, the aim of operation stage is to guarantee that actual performance of menu is consistent with the expected performance.

Finally, menu development corresponds to strategic evaluation since it comprises analysis and improvements. Menu development allows managers to monitor and analyze the menu performance periodically and to make improvement decisions relying on the performance analysis. The improvement decisions may aid establishing a better alignment between the external environment and internal functioning of the firm which means a better strategic fit. In this way, menu development produces feedback to menu planning, pricing and designing. The feedback from menu development stage is also an indicator of deficiencies in the menu operation in terms of deviations from standard procedures. Moreover, the information provided by menu development may indicate to make some adjustments to the objectives of firm.

5. LIMITATIONS AND SUGGESTIONS FOR FUTURE RESEARCH

The conceptual structure of menu management presented in this study provides a tool for hospitality researchers to investigate the menu issues. Researchers can use this conceptual structure as an organizing tool in order to have an overall view of menu management concept and process. For instance, the relationships among menu management issues need to be examined through employing empirical research design. Theoretically, each step of menu management process is an essential preceding step for next steps. Additionally, the efficacy of each menu management issue positively relates to performance of menu management and overall performance of restaurant firm. However, these conceptual relationships need to be empirically investigated.

5. Limitations and suggestions for future research

Although this study has made an effort to conceptually understand the associations between menu management issues and meal experience, there is another need for empirical research investigating the use of menu in enhancing the meal experience. Consequently, some guiding propositions may be derived from the menu management process. First, menu management issues are interrelated. Second, the effectiveness of menu management issues is positively related to the menu performance. Third, the effective management of menu is positively related to the overall performance of restaurant firms. Finally, menu management issues influence the meal experience perceptions of customers.

This is a conceptual study attempting to identify conceptual structure of menu management. Therefore, the major limitation of the study is that it cannot present empirical evidences which are obtained through a specific research design. However, an extensive literature review was conducted in order to reflect the theoretical and empirical insights from the relevant previous studies. Another limitation of the current paper is that the menu management process defined in this study is a general one and ignores the specific changes that may occur in different segments of restaurant industry. That is why an empirical research study which focuses on a comparative analysis of menu management issues in different segments of restaurant industry such as full service versus quick service is needed and such an empirical study will probably produce very useful findings that enhance our understanding of restaurant menus. Additionally, an interesting empirical investigation might be the exploration of differences in menu management issues depending on the different times of consumption such as management of lunch or dinner menus, business-day menus or seasonal menus. In fact, the time when the menu is consumed by customers probably requires different strategies, but because of the space constraints it was not possible to explain those strategies in the current paper. In conclusion, this study fills a gap in the menu literature through offering an integrative conceptual structure of menu management.

REFERENCES

1. Antun and Gustafson, 2005 J.M. Antun, C.M. Gustafson **Menu analysis: design, merchandising and pricing strategies used by successful restaurants and private clubs** J. Nutr. Recipe Menu Dev., 3 (3) (2005), pp. 81–101
2. Ashill and Jobber, 1999 N.J. Ashill, D. Jobber **The impact of environmental uncertainty, decision-maker characteristics and work environment characteristics on the perceived usefulness of marketing information systems (mkis): a conceptual framework** J. Mark. Manag., 15 (1999), pp. 519–540
3. Bayou and Bennett, 1992 M.E. Bayou, L.B. Bennett **Profitability analysis for table service restaurants** Cornell Hotel Restaur. Adm. Q., 33 (2) (1992), pp. 49–55
4. Bernstein et al., 2008 D. Bernstein, M. Ottenfeld, C.L. Witte **A study of consumer attitudes regarding variability of menu offerings in the context of an upscale seafood restaurant** J. Foodserv. Bus. Res., 11 (4) (2008), pp. 398–411
5. Borchgrevink and Ancill, 2003 C.P. Borchgrevink, D.M. Ancill **Beverage control systems and the guests' perspective: standardization and value** J. Hosp. Leis. Mark., 10 (1/2) (2003), pp. 151–159
6. Bowen and Morris, 1995 J.T. Bowen, A.J. Morris **Menu design: can menus sell?** Int. J. Contemp. Hosp. Manag., 7 (4) (1995), pp. 4–9
7. Carmin and Norkus, 1990 J. Carmin, G. Norkus **Pricing strategies for menus: magic or myth?** Cornell Hotel Restaur. Adm. Q., 31 (3) (1990), pp. 7–44
8. Choi et al., 2010 J.G. Choi, B. Lee, J. Mok **An experiment on psychological gaze motion: a re-examination of item selection behavior of restaurant customers** J. Glob. Bus. Technol., 6 (1) (2010), pp. 68–79

9. Cohen et al., 2006 E. Cohen, R. Ghiselli, Z. Schwartz **The effect of loss leader pricing on restaurant menus' product portfolio analysis** J. Foodserv. Bus. Res., 9 (1) (2006), pp. 21–38

10. Davis and Stone, 1998 J.D. Davis, S. Stone **Food and Beverage Management** Butterworth Heinemann, Oxford (1998)

11. Dess and Beard, 1984 G. Dess, D. Beard **Dimensions of organizational task environments** Adm. Sci. Q., 29 (1984), pp. 52–73

12. DiPietro et al., 2004 R.B. DiPietro, M. Roseman, R. Ashley **A study of consumers' response to quick service restaurants' healthy menu items: attitudes versus behaviors** J. Foodserv. Bus. Res., 7 (4) (2004), pp. 59–77

13. Duncan, 1972 R.B. Duncan **Characteristics of organizational environments and perceived environmental uncertainty** Adm. Sci. Q., 17 (1972), pp. 313–327

14. Eves and Dervisi, 2005 A. Eves, P. Dervisi **Experiences of the implementation and operation of hazard analysis and critical control points in the foodservice sector** Int. J. Hosp. Manag., 24 (2005), pp. 3–19

15. Fatimah et al., 2011 U.Z.A.U. Fatimah, H.C. Boo, M. Sambasivan, R. Salleh **Foodservice hygiene factors – the consumer perspective** Int. J. Hosp. Manag., 30 (2011), pp. 38–45

16. Glanz et al., 2007 K. Glanz, K. Resnicow, J. Seymour, K. Hoy, H. Stewart, M. Lyons, J. Goldberg **How major restaurant chains plan their menu: the role of profit, demand, and health** Am. J. Prev. Med., 32 (5) (2007), pp. 383–388

17. Green and Weaver, 2008 Y.N.J. Green, P.A. Weaver **Approaches, techniques, and information technology systems in the restaurants and foodservice industry: a qualitative study in sales forecasting** Int. J. Hosp. Tourism Adm., 9 (2) (2008), pp. 164–191

18. Guéguen et al., 2012 N. Guéguen, C. Jacob, R. Ardiccioni **Effect of watermarks as visual cues for guiding consumer choice: an**

experiment with restaurant menus Int. J. Hosp. Manag., 31 (2) (2012), pp. 617–619

19. Gustafsson, 2004 I. Gustafsson **Culinary arts and meal science – a new scientific research discipline** Food Serv. Technol., 4 (1) (2004), pp. 9–20
20. Hansen et al., 2005 K.V. Hansen, Q. Jensen, I. Gustafsson **The meal experience of a la carte restaurant customers** Scand. J. Hosp. Tourism, 5 (2) (2005), pp. 135–151
21. Hayes and Huffman, 1985 D.K. Hayes, L. Huffman **Menu analysis: a better way** Cornell Hotel Restaur. Adm. Q., 25 (4) (1985), pp. 64–70
22. Hwang and Lorenzon, 2008 J. Hwang, C.L. Lorenzon **Effective nutrition labeling of restaurant menu and pricing of healthy menu** J. Foodserv., 19 (2008), pp. 270–276
23. Iglesias and Guillen, 2002 M.P. Iglesias, M.J.Y. Guillen **The components of total perceived price: an empirical analysis in restaurant services** J. Foodserv. Bus. Res., 5 (1) (2002), pp. 1–22
24. Johns and Kivela, 2001 N. Johns, J. Kivela **Perceptions of the first time restaurant customer** Food Serv. Technol., 1 (2001), pp. 5–11
25. Jones and Mifli, 2001 P. Jones, M. Mifli **Menu development and analysis in UK restaurant chains** Tourism Hosp. Res., 3 (1) (2001), pp. 61–71
26. Jones, 2008 T.A. Jones **Improving hotel budgetary practice – a positive theory model** Int. J. Hosp. Manag., 27 (2008), pp. 529–540
27. Jurkovich, 1974 R. Jurkovich **A core typology of organizational environments** Adm. Sci. Q., 19 (1974), pp. 380–394
28. Kasavana and Smith, 1982 M.L. Kasavana, D.J. Smith **Menu Engineering** Hospitality Publications Inc., Lansing, MI (1982)
29. Kelly et al., 1994 T.J. Kelly, N.M. Kiefer, K. Burdett **A demand-based approach to menu pricing** Cornell Hotel Restaur. Adm. Q., 50 (3) (1994), pp. 383–387

30. Kelly et al., 2009 Kelly, T.J., Kiefer, N.M., Burdett, K., 2009. A Demand-Based Approach to Menu Pricing, Cornell Hos. Q. 50 (3), 383–387.

31. Kincaid and Corsun, 2003 C.S. Kincaid, D.L. Corsun **Are consultants blowing smoke? An empirical test of the impact of menu layout on item sales** Int. J. Contemp. Hosp. Manag., 15 (4) (2003), pp. 226–231

32. Kivela, 1997 J.J. Kivela **Restaurant marketing: selection and segmentation in Hong Kong** Int. J. Contemp. Hosp. Manag., 9 (3) (1997), pp. 116–123

33. Kivela, 2003 J.J. Kivela **Results of a qualitative approach to menu planning using control and experimental groups** J. Foodserv. Bus. Res., 6 (4) (2003), pp. 43–65

34. Kivela et al., 2000 J.J. Kivela, R. Inbakaran, J. Reece **Consumer research in the restaurant environment. Part 3: Analysis, findings and conclusions** Int. J. Contemp. Hosp. Manag., 12 (1) (2000), pp. 13–31

35. Kwong, 2005 L.Y.L. Kwong **The application of menu engineering and design in Asian restaurants** Int. J. Hosp. Manag., 24 (2005), pp. 91–106

36. LeBruto et al., 1995 S. LeBruto, R. Ashley, W. Quain **Menu engineering: a model including labor** FUI Hosp. Rev., 13 (1) (1995), pp. 41–50

37. Lockyer, 2006 T. Lockyer **Would a restaurant menu item by any other name taste as sweet?** FUI Hosp. Rev., 24 (1) (2006), pp. 21–31

38. McCall and Lynn, 2008 M. McCall, A. Lynn **The effects of restaurant menu item description on perceptions of quality, price, and purchase intention** J. Foodserv. Bus. Res., 11 (4) (2008), pp. 439–445

39. Miller, 1980 J.E. Miller **Menu Pricing and Strategy** CBI Publishing, Boston (1980)

40. Miller et al., 1991 J.L. Miller, C.S. McCahon, B.K. Bloss **Food production forecasting with simple time series models** Hosp. Res. J., 14 (3) (1991), pp. 9–21

Chapter 8

Role of Service Environment for Restaurants: The Youth Customers' Perspective

Boo Ho Voon*

Universiti Teknologi MARA, Jalan Meranek, 94300 Kota Saramahan, Sarawak, Malaysia

ABSTRACT

Restaurants are important in generating income for many Asian countries. Therefore, determinants for restaurant service quality need to be identified, especially for the important segments (i.e., youths). This paper examines the importance of service environment (servicescape and human service), food quality and price from the youth's perspective. The types of restaurant investigated were the fine dining, fast food, and food court restaurants. A total of 407 usable questionnaires were analysed. The key determinants on satisfaction and loyalty were human service and price. The impact of servicescape on customer satisfaction and loyalty was relatively smaller, except for fast food restaurants.

KEYWORDS

Service environment; youth; restaurant; youth customer; servicescape

1. INTRODUCTION

The restaurant industry is one of the key contributors for the service sector in many Asian countries. It is relatively competitive in nature with stiff competition within the industry due to an increasing number of new entrants offering differentiated food products and services. On the other hand, the customers are becoming more demanding (Stevens *et al.*, 1995) and their needs are rather diverse. Service has been recognized as one of the effective tools for gaining as well as sustaining the competitive advantage. Nevertheless, achieving superior service for the various types of customers is very challenging due to the nature of the restaurant industry which is characterised by high-contact and people-processing services.

Achieving high level of customer satisfaction for the probable customer loyalty is essential for improving competitiveness of restaurants irrespective of the types. In this respect, it is important to understand the role and relative importance of the numerous key factors which will most probably influence satisfaction and loyalty. This paper specifically investigates the factors, namely Servicescape, Human Service, Food Quality and Price in influencing the youth satisfaction and loyalty in fine dining, fast food and food court restaurants in Malaysia. These restaurants are chosen as they are having significant number of youth customers. The role of service environment is specifically targeted as the role of dining environment, atmosphere and/or tangibles are gaining attention of the service researchers in hospitality.

2. LITERATURE REVIEW AND THEORETICAL FRAMEWORK

Service researchers prefer to measure as well as investigate the relationships among service quality, customer satisfaction and behavioural intentions (e.g. Dabholkar *et al.*, 2000; Qin and Prybutok, 2008; Zeithaml *et al.*, 1996). Generally, the customer-perceived service quality is found to be positively related to behavioral intentions and loyalty. There are also researchers (e.g. Dabholkar *et al.*, 2000; Brady and Robertson, 2001; Tam, 2004) who found that customer satisfaction mediates the effect of service quality on behavioral intentions.

On the other hand, Wong (2004) found that service quality positively would affect emotional satisfaction that in turn would have positive effect on both customer loyalty and relationship quality. Additional support for the links between service quality, customer satisfaction and loyalty has been provided by Tsoukatos and Rand (2006) who also differentiated between emotional loyalty and behavioural loyalty. Emotional loyalty was said to mediate the effect of satisfaction on behavioural loyalty. The importance of service quality on customer loyalty intentions was highlighted by Venetis and Ghauri (2004) who found that service quality affected customers' intentions to stay in a relationship with a service provider. Service quality also may affect loyalty through the mediating variables of trust and commitment (Caceres and Paparoidamis, 2007).

The servicescape or mainly the tangibles of the restaurants has also been found to affect customers' evaluations of service quality as well as post-purchase behavior (e.g. Kim and Moon, 2009). In fact, Bitner (1992) had presented a comprehensive model for understanding how the physical surroundings of a service organization may influence individual behaviors of customers. The dimensions include ambient conditions such as noise, music, scent, lighting and temperature, spatial layout which includes the arrangement of furniture and equipment, and signs, symbols and artifacts such as décor and signage. Bitner (1992) argued that the intangible nature of services may lead consumers to make cognitive assessments of service quality based on extrinsic cues found in the servicescape. This physical environment (Wall and Berry, 2007) may affect customers' physical comfort and movement. On the contrary, Voon *et al.* (2007) found that servicescape had relatively small impact on the satisfaction and loyalty of the restaurant customers. As such, this paper aims to know the role the more comprehensive service environment from the youth customer's perspective.

In addition, the role of food quality is also recognized as one of the key determinant for customer satisfaction and loyalty. Good quality of food and beverages can be instrumental in satisfying and retaining the youth customers who most probably enjoy eating delicious food. There are many Malaysian youth customers who prefer hot and spicy food available in fast food and food courts. Researchers have found that the quality of menu items affects customers' intention to dine at the restaurant (e.g. Qu, 1997; Kivela *et al.*, 2000; Soriano, 2002; Zopiatis and Pribic, 2007).

Pricing or the perceived-value of the restaurants is another factor that has been studied in relation to customer satisfaction and patronage intentions. Monetary and time costs negatively affect perceived value while perceived service quality leads to higher levels of perceived value. Furthermore, post-purchase intentions were more strongly influenced by perceived value (Tam, 2004). Qin and Prybutok (2008) also investigated the role of price/value in determining customer satisfaction for fast-food restaurants but did not find it to be of significance. This is however believed to be due to the nature of fast-food restaurants in which price is relatively low and therefore not of central importance to consumers compared to other types of establishments such as fine-dining restaurants. Nevertheless, to the youth who are not economically strong may find that price is important to attract them to the restaurants.

Undoubtedly, many findings suggest that servicescape, human-related service, price and food quality have influences on customer satisfaction and loyalty. However, some studies have only considered human service quality, price and food quality without including the tangibles. On the other hand, scant attention on the youth market. This study therefore contributes to existing research by comparing the effects of all four factors on satisfaction and loyalty. The proposed relationships are shown in Fig. 1.

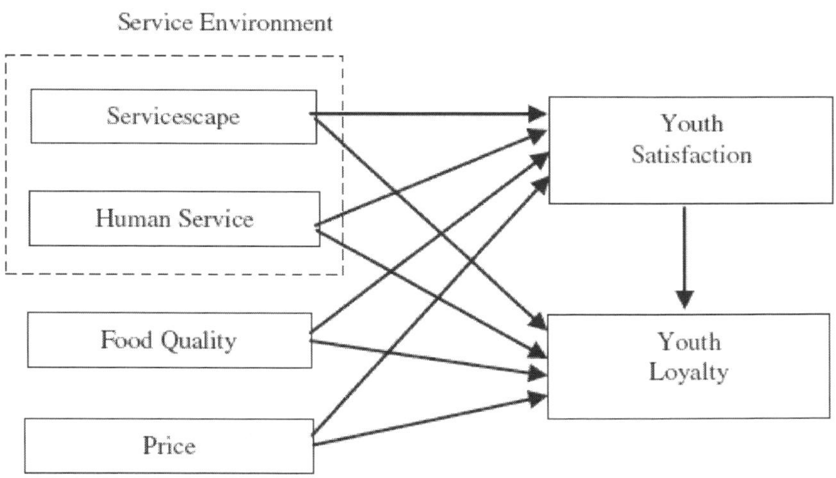

Figure. 1. Relationships among servicescape, human service, food quality, price, youth satisfaction and youth loyalty

3. METHODOLOGY

The focus of this study is mainly on the role of restaurant service environment (servicescape and human service) in the different types of restaurants, from the Malaysian youths' perspective. The youth customers who had recently dined at fine dining, fast food and food court restaurants in the previous three months were invited to answer the structured questionnaires. The fine dining restaurants were those modern restaurants with nice and comfortable settings found in hotels, fast food restaurants were the Western-style fast food outlets (e.g. KFC, McDonald's, Pizza Hut and Chicago 7) whereas food court restaurants were mainly the food establishments (consist of many food stalls) operated by the locals serving mainly the popular local food and drinks.

The samples were selected using quota sampling. The key controlled characteristics were gender, age, race and education. Youths were customers who aged from 15 to 25 years old. The total number of usable questionnaires was 407 (Fine Dining, 145; Fast Food, 147 and Food Court, 117). The characteristics of the respondents who participated in the survey are shown in Table 1.

The multi-item measures for human service, price and food quality were adapted from the scales of Qin and Prybutok (2008). During exploratory factor analysis using pilot testing data, one of the dimensions for human service (i.e, responsiveness) was excluded due to significant cross-loading with the other service dimensions. The servicescape items were adapted from Kim and Moon (2009) and Bitner (1992) while the scales of Cronin and Taylor (1992) and Boulding et al. (1993) were adapted for measuring customer satisfaction and loyalty respectively. The summarized items for the various measurement scales with their respective operational items are as follows:

Environment includes things and people around us in life. As such, the *Service Environment* encompasses the servicescape (physical environment) and the human service aspects of the restaurants. The items for *Servicescape* include the attractiveness of the physical facilities, availability of parking space, cleanliness, temperature, music, colour scheme, smell/scent, labels/signage and design/layout). This concerns mainly the purposively created environment to serve the customers.

Human service (Assurance: well-dressed and appear neat, wear sanitary gloves and hair net, friendly, polite, knowledgeable, can be trusted; Reliability: promised service as promised, helpful, made me feel confident, charged accurately, on-schedule service; Empathy: individual attention, convenient operating hours, complete packaged food, sauces/spices always available).

Food quality (clean, healthy, fresh and has a variety)

Price (competitive price, good value for money, willing to give special prices)

Youth satisfaction (overall satisfaction, quality of service experience, meet expectation)

Youth loyalty (repurchase intention and recommend to others)

These items were measured using the 7-point Likert scale (1=Strong Disagree, 7=Strong Agree) in the structured questionnaire. Reliability analysis (Cronbach's alpha coefficients for measures and internal consistency method for items using item-to-total correlations) and multiple regressions were done on the respective measures to understand the goodness of measures, relative importance and relationships.

4. FINDINGS AND DISCUSSIONS

The results of the reliability tests are shown in Table 2. All the six multi-item variables (Servicescape, Human Service, Price, Food Quality, Youth Satisfaction and Loyalty) are found to be reliable as their Cronbach's alpha coefficients are close to or higher than the required level of 0.70 (Nunnally and Bernstein, 1994). The items are also showing acceptable levels of item-to-total correlation results.

Table 1. Respondent Profile

Demographic variables		Fine Dining (n=145) %	Fast Food (n=147) %	Food Court (n=117) %
Gender:	Male	42.1	42.2	35.9
	Female	57.9	57.8	64.1
Age:	15-20	39.3	13.4	17.1
	21-25	60.7	86.6	82.9
Race:	Malays	61.4	49.2	41.9
	Dayaks	20.7	21.4	30.8
	Chinese	12.0	12.3	12.8
	Others	5.9	17.1	14.6
Education:	Diploma or lower	66.9	32.6	47.9
	Degree	30.3	55.1	40.2
	Professional	4.0	12.1	12.0

The findings from the different types of restaurants indicate that the youth customers who participated in the survey were generally satisfied with the variables involved for the restaurants. However, they are particularly more satisfied with Food Quality and the Servicescape aspects of the restaurants as these factors received relatively higher mean scores for the restaurants. This indicates that the restaurants patronized by the respondents in this study performed better in terms of the quality of their food and environment. The relatively lower mean scores were Human Service and Price. This implies that the factor which least satisfies the restaurants' youth patrons is the human-related service provided by the different types of restaurants.

The relative importance of the four factors in influencing the youths' satisfaction and loyalty were also investigated. The findings in Table 3 show that, for Fine Dining, these four determinants explained 67% of the variance in customer satisfaction, Fast Food (58%) and Food Court (52%). For all the types of restaurants, Human Service is the most important factor (Rank 1) which exerted a positive and significant influence on their satisfaction.

Table 2. Reliability and descriptive statistics for variables

Variables (no. of items)	Fine Dining			Fast Food			Food Court		
	Alpha	Mean	S.D.	Alpha	Mean	S.D.	Alpha	Mean	S.D
Servicescape (9)	0.83	4.68	0.91	0.87	4.87	0.95	0.86	5.06	0.84
Human Service (15)	0.93	4.54	1.06	0.87	4.77	0.77	0.90	5.04	0.80
Food Quality (4)	0.79	4.76	1.01	0.62	4.89	0.85	0.72	5.12	0.89
Price (3)	0.77	4.54	1.19	0.74	4.69	1.19	0.88	5.21	1.19
Customer Satisfaction (3)	0.81	4.70	1.12	0.78	4.95	1.07	0.88	4.89	1.26
Customer Loyalty (2)	0.87	4.78	1.30	0.73	5.01	1.24	0.62	5.05	1.19

Though the Human Service dimension of the service environment contributed much to the youth satisfaction, the Servicescape did not seem to impact their satisfaction conclusively. The results in Table 3 indicate that for Fine Dining and Food Court, Servicescape does not seem to have significant influence on the youth satisfaction. They should have their pre-determined expectations of these restaurants. They expect the former to have very good environment where as the latter most probably less in terms of conducive and nice environment. But, they could be exceptional cases where the youths may prefer the crowded and environment. However, for the Fast Food sample, Servicescape has significant effect on youth satisfaction and ranked second.

Price or the perceived value for money was generally found to be the most important factor in determining the youth loyalty, followed by Human Service. Servicescape and Food Quality were relatively less important and not significant (at 0.05 level) in influencing their loyalty. This finding is in contrast to prior research by Kim and Moon (2009). Most probably, there could be other necessary moderating factor(s) for these relationships. In fact, Harris and Ezeh (2008) argued that certain variables related to service environment in fact had significant association with loyalty when moderating factors were included. This suggests that the relationship between service environment, satisfaction and loyalty may not be a direct one but is in fact more complex, possibly involving mediating and/or moderating factors. Emotional experience was found to be one of them (Kim and Moon, 2009). Hence, the findings in this study can still be further researched In order to arrive at more conclusive findings. Further validations are necessary.

The critical role of Human Service in determining the youth satisfaction and loyalty is generally clear and managers should emphasize the quality of human service at all times across different types of restaurants. Nevertheless, the role of Servicescape is still questionable. The important attributes of Servicescape should be explored. Since

Servicescape ranked second in influencing satisfaction, the relative importance of its attributed was explored and the results are shown in Table 4. The significant attributes (at 0.05 level) are namely: 'Appropriate labels and signage', 'Has a good design and layout' as well as 'Parking space always available.'

Table 3. Relative importance of the variables on youth satisfaction. *(Significant at 0.05 level)

Independent	Satisfaction								
	Fine Dining			Fast Food			Food Court		
Variables	B	Sig.	Rank	β	Sig.	Rank	B	Sig.	Rank
Servicescape	-0.033	0.703	4	0.243*	0.004	2	0.069	0.452	3
Human Service	0.487*	0.000	1	0.375*	0.000	1	0.598*	0.000	1
Food Quality	0.214*	0.003	3	0.061	0.386	4	0.032	0.700	4
Price	0.220*	0.009	2	0.190*	0.017	3	0.082	0.345	2
R^2	0.67			0.58			0.52		

Table 4. Relative importance of the variables on youth loyalty. *(Significant at 0.05 level)

Independent	Loyalty								
	Fine Dining			Fast Food			Food Court		
Variables	B	Sig.	Rank	β	Sig.	Rank	B	Sig.	Rank
Servicescape	0.158	0.090	3	-0.06	0.526	4	0.056	0.463	3
Human Service	0.280*	0.015	2	0.509*	0.000	1	0.398*	0.000	2
Food Quality	0.140	0.071	4	0.139	0.083	3	0.045	0.518	4
Price/Value	0.293*	0.001	1	0.150	0.096	2	0.435*	0.000	1
R^2	0.61			0.46			0.67		

Table 5. Relative importance of servicescape attributes. *(Significant at 0.05 level)

	Fast Food		
Attributes	β	Sig.	Rank
Appealing physical facilities	0.123	0.178	4
Parking space always available *	0.182*	0.018	3
Clean dining area	0.024	0.804	8
Appropriate temperature	0.036	0.660	7
Has a nice music	0.103	0.213	5
Has appropriate colour scheme	0.067	0.464	6
Has a nice smell (scent)	-0.078	0.420	9
Appropriate labels and signage *	0.267*	0.002	1
Has a good design and layout *	0.207*	0.018	2

5. CONCLUSIONS AND FUTURE RESEARCH

This purpose of this study on fine dining, fast food and food court restaurants in Malaysia was to identify and understand the relative importance of four determinants of customer satisfaction and loyalty, namely, Servicescape, Human Service, Food Quality and Price. Overall, it was found that Service Environment (Servicescape and Human Service) played a critical role in determining the youth satisfaction and loyalty.

The results indicated that customers were generally satisfied with the restaurants and the key determinants on their satisfaction and loyalty were mainly Human Service and Price. The former was found to be the most important across all the three types of restaurants investigated. However, the impact of servicescape on customer satisfaction and loyalty was found to be relatively smaller, except for the fast food restaurants.

The managerial implication for restaurant managers and operators is that careful attention and proper resource allocation should be given to deliver high levels of Human Service in order to delight customers. The customer-contact service employees must be appropriately trained and monitored in terms of their skills knowledge, attitudes and relations (SKAR). The employees need to be specifically empathetic or caring, reliable, responsive and courteous. They must be motivated and emotionally stable at all times.

The findings suggest that merely emphasizing the physical environment of the restaurants will not necessarily improve the youth satisfaction and loyalty if the quality of service provided by the restaurant employees (Human Service) is lacking. The intended Service Environment includes Human Service and Servicescape. As such, a more comprehensive enhancement that is based on Service Environment will be more strategic for winning their hearts. Nevertheless, more detailed investigations with more representative samples across different cultural backgrounds will still be needed. Future studies can examine the youth customers of other cultural settings (ethnical, national and international) in order to validate the current findings. The role of potential moderating factors such as emotional experience should be considered. Understanding the nature of relationships among the variables will add more insights for managers. The relationships or consequences could non-linear or indirect. Structural equation modeling (SEM) for examining the structural relationships of the various variables involved will be useful for the managers to allocate resources strategically to serve the youths.

ACKNOWLEDGMENT

The author wish to thank the research assistants and undergraduates who helped in the data collection and preparation processes of the research project. The paper has been slightly revised from the original published.

REFERENCES

1. Bitner, M.J. (1992). Servicescapes: The impact of physical surroundings on customers and employees. *Journal of Marketing*, 56, 57-71.

2. Boulding, W., Kalra, A., Staelin, R. & Zeithaml, V. (1993). A dynamic process model of service quality: from expectations to behavioral intentions. *Journal of Marketing Research*, 30 (February), 7-27.

3. Brady, M.K. and Robertson, C.J. (2001). Searching for a consensus on the antecedent role of service quality and satisfaction: An exploratory cross-national study. *Journal of Business Research*, 51, 53-60.

4. Caceres, R.C. and Paparoidamis, N.G. (2007). Service quality, relationship satisfaction, trust, commitment and business-to-business loyalty. *European Journal of Marketing*, 41(7/8), 836-867.

5. Cronin, J.J. and Taylor, S.A. (1992). Measuring service quality: A reexamination and extension. Journal of Marketing, 56(3), 55-68.

6. Dabholkar, P.A., Shepherd, C.D. and Thorpe, D.I. (2000). A comprehensive framework for service quality: An investigation of critical conceptual and measurement issues through a longitudinal study. *Journal of Retailing*, 76(2), 139-173.

7. Harris, L.C. and Ezeh, C. (2008). Servicescape and loyalty intentions: An empirical investigation. *European Journal of Marketing*, 42(3/4), 390-422.

8. Keillor, B.D., Lewison, D., Hult, G.T.M. and Hauser, W. (2007). The service encounter in a multi-national context. *Journal of Services Marketing*, 21(6), 451-461.

9. Kim, W.G. and Moon, Y.J. (2009). Customers' cognitive, emotional, and actionable response to the servicescape: A test of the moderating effect of the restaurant type. *International Journal of Hospitality Management*, 28, 144-156.

10. Kivela, J., Inbakaran, R. and Reece, J. (2000). Consumer research in the restaurant environment. Part 3: analysis, findings and conclusions. *International Journal of Contemporary Hospitality Management*, 12(1), 13-30.

11. Nunnally, J.C. and Bernstein, I.H. (1994). *Psychometric Theory*, 3rd Ed. New York: McGraw-Hill, Inc.

12. Qin, H. and Prybutok, V.R. (2008). Determinants of customer-perceived service quality in fast-food restaurants and their relationship to customer satisfaction and behavioral intentions. *The Quality Management Journal*, 15(2), 35-50.

13. Qu, H. (1997). Determinant factors and choice intention for Chinese restaurant dining: A Multivariate approach. *Journal of Restaurant & Foodservice Marketing*, 2(2), 35-49.

14. Reimer, A. and Kuehn, R. (2005). The impact of servicescape on quality perception. *European Journal of Marketing*, 39(7/8), 785-808.

15. Stevens, P., Knutson, B. and Patton, M. (1995). DINESERV: A tool for measuring service quality in restaurants. *Cornell Hotel and Restaurant Administration Quarterly,* 36 (2), 56-60.

16. Soriano, D.R. (2002). Customers' expectations factors in restaurants: The situation in Spain. *International Journal of Quality & Reliability Management*, 19(8/9), 1055-1067.

17. Tam, J.L. (2004). Customer satisfaction, service quality and perceived value: An integrative model. *Journal of Marketing Management*, 20, 897-917.

18. Tsoukatos, E. and Rand, G.K. (2006). Path analysis of perceived service quality, satisfaction and loyalty in Greek insurance. *Managing Service Quality*, 16(5), 501-519.

19. Venetis, K.A. and Ghauri, P.N. (2004). Service quality and customer retention: building long-term relationships. *European Journal of Marketing*, 38(11/12), 1577-1598.

20. Voon, B.H. (2011). Service environments of restaurants: Findings from the youth customers, *Journal of Asian Behavioural Studies*, 1(2), 45-56.

21. Voon, B.H., Kueh, K., Unggah, L. And Chali, R. (2009). Customers' perception of restaurant service quality: Evidence from Malaysia. *Social and Management Research Journal*, 6(1), 33-50.

22. Voon, B.H., Chitra, K. and Kueh, K. (2009). Factors influencing customer satisfaction and loyalty in the restaurant industry: Findings from Malaysia and India. *Proceedings of Asia Pacific Marketing Conference*, Kuching, Malaysia.

23. Wall, E. and Berry, L. (2007). The combined effects of the physical environment and employee behaviour on customer perception of restaurant service quality. Cornell Hotel and Restaurant Administration Quarterly, 48(1), 59-69.

24. Wong, A. (2004). The role of emotional satisfaction in service encounters. *Managing Service Quality*, 14(5), 365-376.

25. Zeithaml, V.A., Berry, L.L. and Parasuraman, A. (1996). The behavioral consequences of service quality. *Journal of Marketing*, 60 (April), 31-46.

26. Zopiatis, A. and Pribic, J. (2007). College students' dining expectations in Cyprus. *British Food Journal, 109(10), 765-776.*

Index